Russian Views of the Transition in the Rural Sector

Structures, Policy Outcomes, and Adaptive Responses

L. Alexander Norsworthy, Editor

Environmentally and Socially Sustainable Development
Europe and Central Asia Region
The World Bank
Washington, DC

© 2000 The International Bank for Reconstruction
and Development/THE WORLD BANK
1818 H Street, N.W.
Washington, D.C. 20433

All rights reserved
Manufactured in the United States of America
First printing June 2000
1 2 3 4 5   04 03 02 01 00

The findings, interpretations, and conclusions expressed in this paper are entirely those of the author(s) and should not be attributed in any manner to the World Bank, to its affiliated organizations, or to members of its Board of Executive Directors or the countries they represent. The World Bank does not guarantee the accuracy of the data included in this publication and accepts no responsibility for any consequence of their use.

The material in this publication is copyrighted. The World Bank encourages dissemination of its work and will normally grant permission to reproduce portions of the work promptly.

Permission to *photocopy* items for internal or personal use, for the internal or personal use of specific clients, or for educational classroom use is granted by the World Bank, provided that the appropriate fee is paid directly to the Copyright Clearance Center, Inc., 222 Rosewood Drive, Danvers, MA 01923, USA; telephone 978-750-8400, fax 978-750-4470. Please contact the Copyright Clearance Center before photocopying items.

For permission to *reprint* individual articles or chapters, please fax a request with complete information to the Republication Department, Copyright Clearance Center, fax 978-750-4470.

All other queries on rights and licenses should be addressed to the Office of the Publisher, World Bank, at the address above or faxed to 202-522-2422.

Cover photo by Thomas R. Ward, Consultant, ECSSD

Library of Congress Cataloging-in-Publication Data

Russian views of the transition in the rural sector : structures, policy outcomes, and adaptive responses / L. Alexander Norsworthy, editor.
    p. cm.
A collection of articles translated from Russian.
Includes bibliographical references.
ISBN 0-8213-4765-9
    1. Agriculture and state—Russia (Federation) 2. Russia (Federation)—Rural conditions. I. Norsworthy, L. Alexander, 1962–   II. World Bank.

HD1995.15.Z8 R87 2000
333.76'1947—dc21

00–032476

CONTENTS

Acknowledgments v

Abstract vii

Preface ix

Abstracts of Russian Articles xi

L. Alexander Norsworthy and Olga Paluba "Impact of the Transition: Approaches and Findings" 1

Maria Amelina "Rural Interactions in the Post-Soviet Era" 16

Vasily Uzun "Agrarian Reform in Russia in the 1990s: Objectives, Mechanisms, and Problems" 25

Renata Yanbykh "Driving Forces in Russian Agrarian Policy in the 1990s" 42

Vladimir Bogdanovsky "Land Reform: Expectations and Social Consequences" 57

Eugenia Serova "Public Opinion concerning Russia's Agrarian Reforms" 67

Zemfira I. Kalugina "Paradoxes of Agrarian Reform in Russia" 87

Eugenia Serova "Reform and Economic Behavior in Russian Agriculture" 103

Zemfira I. Kalugina "Survival Strategies of Enterprises and Families in the Contemporary Russian Countryside" 118

Olga V. Artemova "Changes in the Everyday Activities of Rural Women in Russia from the 1970s to the 1990s" 132

Grigory Ioffe and Tatiana Nefedova "The Environs of Russian Cities: A Case Study of Moscow" 145

Galina Vitkovskaya "Adaptation and Integration of Forced Migrants in Russia" 167

Sergei A. Panarin "The Rural Economy of the Tunka Valley in a Time of Transition and Crisis" 188

L. Alexander Norsworthy and Alexander Feshenko "Studying the Political Economy of the Rural Transition" 202

ACKNOWLEDGMENTS

To arrange for the publication of a compendium of Russian articles on different political, economic and social aspects of rural reform in Russian was a difficult and time consuming task. This effort was undertaken to demonstrate the need for a multidisciplinary approach to analyzing a decade of reforms. This activity required the support of many World Bank project leaders and managers in the Environmentally and Socially Sustainable Development Department of the Europe and Central Asia Region at the World Bank. The gracious cooperation and contributions of Russian experts from Moscow to Siberia, and within the World Bank itself made this concept a reality.

I am particularly grateful for the feedback and guidance from Ayse Kudat, Thomas Blinkhorn and Stan Peabody of the Social Development Team. Also the suggestions from Laura Tuck, John Nash and Csaba Csaki of the Rural Development team, and Kevin Cleaver, the Department Director.

Maria Amelina's work in the Development Economics Research Group closely matched many of my original interests in exploring what the Russians themselves had to say about the reform experience. Christian Forster of the US Department of Agriculture was also an important sounding board for what this compendium was trying to accomplish.

A special thanks is to Frank Laczko of IOM for allowing the reprint of the Vitkovskaya article on Adaptation Strategies. I am also grateful for recommendations from the publications coordinator of the Social Development Anchor Bonnie Bradford and from Paola Scalabrin in the Office of the Publisher.

The papers in this compendium were obtained as the original work of the authors in English and reflect the opinions of the authors, not those of the World Bank Group. Any errors or omissions must be attributed to the editor, not the contributors

ABSTRACT

This compendium of papers by Russian experts on the rural sector seeks to explore the institutions, policy outcomes and adaptive responses of enterprises and households to restructuring of the rural sector. The importance of hosuehold plots has increased dramatically but these production units are still dependent on the large successors to the collective and state farms. There are some hopeful signs that the markets are responding to the reforms and producing more efficiently, but this is offset by the uneven implementation of new farm and enterprise laws accross the different regions. Because of demgraphic shifts and the change in land markets, the boundaries between rural and urban are blurred particularly in large metropolitan areas. Rural women continue to bear a disproportionate burden of household work, and the opportunities for migrants moving to rural communities to match or exceed their previous standards of living are constrained. This is due to the importance of social networks of households and enterprises and access to scarce inputs in rural areas.

PREFACE

This compendium is not a typical World Bank publication. Most of the contributing authors are Russian academicians and policy specialists living in Russia. Their approach to evaluating the effects of the transition in that sector differ substantially from the World Bank's.

Many terms such as "land ownership" and "private" or "independent farm" do not have the same meanings among the Russian experts themselves or among rural households. Definitions differ even more markedly from those employed in Western countries. The boundaries between the firm and the household and between the government and the large rural enterprises remain blurred. The boundaries between government and non-government are not always apparent in the attached volume. This is indicative of the confusion which exists regarding causes of the deterioration of Russian agriculture and solutions. Also, the different papers do not present comparable economic and social indicators, thus posing an obstacle to synthesizing and integrating their research findings with those of the Bank. Despite these at times contradictory views of what has happened in Russia and why, there is value in reviewing these interpretations and policy prescriptions. The intent of the compendium is to raise awareness of Russian perspectives on the transition in rural areas rather than to advocate specific development interventions from a World Bank perspective.

The World Bank perspective has been published elsewhere and circulated widely. The policy prescriptions and recommendations regarding the agricultural sector in Russia presented by the World Bank have been ignored in Russia. Most of the authors of the articles admit the failure of Russian agriculture in the "transition," but provide a different set of solutions to those most often seen in the West. We hope that this will stimulate a debate that can contribute to moving the sector forward. We are grateful to the Russian specialists for their contributions to this activity and hope to continue this knowledge partnership with them and their institutions in the future.

> Kevin Cleaver,
> Sector Director,
> Environmentally and Socially
>   Sustainable Development
> Europe and Central Asia Region

ABSTRACTS OF RUSSIAN PAPERS

Vasily Uzun "Agrarian Reform in Russia in the 1990s: Objectives, Mechanisms, and Problems"

The constitutional amendments on landownership approved by the Russian legislature in 1990 secured the right to private ownership of land, for the first time since the 1917 revolution. The amendments and the implementing directives and regulations that followed provided for the introduction of private ownership of land and for transfer of the property held by collective and state-owned farms to the ownership of their members. But agreement on the details, and even some of the principles, of agrarian reform was not unanimous, and many practical and political hurdles stood in the way. Not least of the problems was that instructions from the central government were not necessarily carried out on lower levels.

Agrarian reform was needed—episodic drives and interventions had not been able to remedy the fundamental inefficiency of Russian agriculture. After considering many options, including those used in other countries, policymakers finally selected a course that was politically and practically feasible for Russia. Changes in the organization of agriculture took place, but private agriculture failed to blossom immediately.

There are many barriers in the way of the (still incomplete) reform. Some are psychological, such as the dearth of entrepreneurship, which had been suppressed for decades; some are rooted in competing political and institutional interests. The disparities between prices for agricultural products and for inputs and manufactured goods condemn many farmers and enterprises to a constant struggle against losses, debt, and failure.

Renata Yanbykh "Driving Forces in Russian Agrarian Policy in the 1990s"

A large number of institutions and organizations—governmental and nongovernmental—make or influence Russia's agricultural and food policy. This article surveys the legislative and government bodies responsible for elaborating agricultural policy, outlines the agrarian programs of the main political parties in Russia, and describes the emergence of business associations in the agricultural and food sector. Because of the influence that research and consulting organizations have on agrarian policy, several of the most important institutes are described.

Vladimir Bogdanovsky "Land Reform: Expectations and Social Consequences"

A perhaps naïve view of land reform saw the simple act of distributing land as the remedy for the economic and social defects that have haunted Russian agriculture. The results of the reform, however, have been mixed; food production has not increased as expected, and social investment and living standards in the countryside have deteriorated.

Surveys conducted in 1990 and 1995 of farmers and agricultural employees in four regions of the Russian Federation show that even modest expectations of improvements resulting from land reform have not been met. One consequence has been a loss of faith in the reforms and in the government.

Despite the de facto growth of private agriculture, much of it on subsidiary plots still legally attached to the collective farm, many farmers oppose formal private landownership, are not interested in it, or are simply uninformed or confused. A significant proportion prefers to continue working on their subsidiary plots within a collective system.

Eugenia Serova "Public Opinion concerning Russia's Agricultural Reforms"

Eight years into the post-Soviet era, reform of the agricultural sector is still a matter of contention, and the reform itself has had only modest results. To assess the degree to which government leaders, managers, and influential experts support the reforms, a survey of 500 members of the "agricultural establishment" in Moscow and in seven other *oblasts* (regions) was conducted in 1999. The results were broken down by profession, age group, political inclination, and region. The overall findings confirm a lack of consensus on the concept of the reforms—a circumstance that distinguishes Russia from the Central and Eastern European countries and that may be a significant constraint on the transition in agriculture.

The need for land reform was generally accepted, but there was much more skepticism about privatization of the processing chain, perhaps because of fear of monopsony that would impair farmers' interests. Opinion on actual implementation of the reforms varied across the sample, often in a predictable way, with younger experts and the political right being more optimistic about the effectiveness of the measures than were older groups and the left. The majority of the experts thought that cooperatives would be the dominant form of production unit for some time to come, but household plots and large commercial farms were also seen as contenders. Only about 12 percent of the sample thought family farms would dominate.

The questionnaire included sections on the social and political aspects of the reforms and on the future of Russian agriculture. About 84 percent of the sample believed that the reforms had led to a worsening of living standards in rural areas. Opinion about the future varied, but most respondents said that recovery in agriculture would not begin within the next 10 years. Few saw any prospects that Russia will become a net grain exporter any time soon.

The widespread pessimism about the implementation of the reforms, if not about the underlying principles, points to a lack of confidence that could turn into a self-fulfilling prophecy. Successful, replicable examples could offer a way of breaking through the inertia and strengthening support for the reforms and faith in the future.

Zemfira Kalugina "Paradoxes of Agrarian Reform in Russia"

Although the reforms of the 1990s were supposed to foster the growth of private farms, the reform period has in fact seen a growth in the share of output from household farms, run by rural dwellers who are still members of a collective agrarian enterprise. Meanwhile, the share of collective enterprises is declining. The author's survey conducted in the Novosibirsk area in 1990–97 indicated, among other findings, that few operators of household farms see the possibility of becoming independent of the collective enterprises. Among the reasons are their need for support from the collective enterprise and the likelihood of a heavier tax burden. In addition, the price structure is unfavorable to agriculture, and the social and marketing infrastructure needed by independent farmers is not yet in place.

The reforms have given rise to several unexpected outcomes, or paradoxes: the expansion of small-scale commodity production; inefficiency in private sector agriculture; the destruction of motivation to work, and the worsening of rural living conditions as social infrastructure deteriorates. The social costs of top-down reform have led to disappointment and loss of confidence. The way out lies through greater attention to the real situation in the countryside and greater respect for rural people's preference for collective forms of organization.

Eugenia Serova, "Reform and Economic Behavior in Russian Agriculture"

Despite numerous economic, political, and psychological handicaps, economic reform in Russian has made possible the emergence of new institutions and behaviors in the agricultural sector.

For the first time, farm managers are seeking to maximize profits rather than to fulfill the state plan, leading to changes in the structure of output. Trends in production of grain, sunflower seeds, and flax fiber illustrate producers' responsiveness to price signals.

Various intermediaries, including private firms and individuals, are moving into the marketing of agricultural products as state purchases decline in importance. Financial discipline is improving with the end of Soviet-era soft loans that were often effectively written off. Banks are now providing agricultural credit, and lenders expect loans to be repaid. A survey of five regions shows that farms of all types, as well as processors, have better repayment records than suppliers and regional governments. Lenders have reacted by extending increased credits to the more reliable borrowers and cutting back on loans to the other organizations.

Problems in the sector include the recent economic crisis, which undermines many of the positive developments described in this paper; the persistence of barter; and the incomplete and unsystematic implementation of the reforms.

A 1998 survey of four selected agricultural enterprises in Novosibirsk *oblast* (region) investigated the enterprises' strategies for adapting to the economic transition from a planned economy to a market system. The study examined how social services, the labor force, and enterprises' economic health fared under different circumstances and strategies and analyzed households' ways of coping with economic change. Four main models of adaptation were identified: active participation in the market; conformity and compensatory activity; the "mimicry," or survival, strategy resorted to by firms short of resources; and passivity (destructive adaptation).

Zemfira I. Kalugina" Survival Strategies of Enterprises and Families in the Contemporary Russian Countryside"

A 1998 survey of four selected agricultural enterprises in Novosibirsk *oblast* (region) investigated the enterprises' strategies for adapting to the economic transition from a planned economy to a market system. The study examined how social services, the labor force, and enterprises' economic health fared under different circumstances and strategies and analyzed households' ways of coping with economic change. Four main models of adaptation were identified: active participation in the market; conformity and compensatory activity; the "mimicry," or survival, strategy resorted to by firms short of resources; and passivity (destructive adaptation).

Olga V. Artemova "Changes in the Everyday Activities of Rural Women in Russia from the 1970s to the 1990s."

A longitudinal time-budget study of rural households in western Siberia provides insights into how the recent reforms have affected the daily lives of agricultural workers and other rural residents. The workload, after reaching a physiological limit in the late 1980s, has declined somewhat, but women still spend more hours working than do men and have less free time. Both men and women are putting in more time on household plots and duties as work outside the household becomes less available, less remunerative, or less satisfying. Shortages of money, difficulties in obtaining provisions, and the deterioration of services in the reform period have led to greater reliance on the household's own resources and on the family as a support mechanism.

Rural women, especially agricultural workers, are generally overworked. Because of women's role in reproduction, family health, and early education, this situation could have detrimental effects on the next generation. A striking finding is that women find little satisfaction in much of their work and feel a lack of free time for true leisure activities or the things they would most like to do.

Grigory Ioffe and Tatiana Nefedova "The Environs of Russian Cities: A Case Study of Moscow"

The fringes of the largest Russian cities—Moscow and St. Petersburg—differ in essential ways from Western, especially American, suburbs. Among the roots of the peculiar features of Russian urban margins are the general desire to live in or near the city (leading to great differences in population density, a clear distinction between city and country, and the depopulation of rural areas); the hankering of urbanites for recreational second dwellings in rural areas, from elaborate *dachas* and country "cottages" to shacks on orchard plots; and the significance of agriculture, including small-scale farming, in the vicinity of cities.

Land-use patterns reflect and provide insights into current social, economic, and political changes. In Moscow city and *oblast* (region), privatization of land and the demand for second dwellings in exurbia (fueled by the relative wealth of Muscovites) have created a booming land and real estate market. Agricultural land near the city is being lost to other uses but perhaps not as much in reality as on paper, as private recreational use includes labor-intensive farming by urbanites. Productivity of agricultural land is, with a few exceptions, higher close to the city than on the periphery of the oblast, partly because of the higher labor intensity. The "normative" prices set by the oblast for tax purposes, while far below market prices, take location into account, a departure from Soviet-era thinking.

Moscow oblast is asserting its independence from the city regarding land use. Local administrations have become de facto owners and sellers of former state land, making them significant beneficiaries of land sales.

State and collective farms, nominally transformed into joint-stock companies, suffered from the crises of the 1990s but still employ about 200,000 people. The tiny subsidiary household farms account for about one quarter of total agricultural output in Moscow oblast, and their production increased by one third between 1991 and 1996. Private farming still contributes only about 1.2 percent of agricultural output, but the sector has grown rapidly since 1990, and there are now about 6,500 private farmers in the oblast.

Galina Vitkovskaya "Adaptation and Integration of Forced Migrants in Russia"

Forced migrants from the former republics of the U.S.S.R. and from troubled zones in Russia have, on average, higher educational levels and professional qualifications than the Russian population as a whole. Their skills and their leanings toward entrepreneurship and independent action are potential assets for the country. The consequence of current policy, however, is that migrants settle in areas that are remote, rural, and depressed. Regular surveys, summarized in this paper, indicate that this policy may have negative consequences in the near future.

Migrants in villages do have the advantage of relatively better housing and access to garden plots, and their resettlement in the countryside injects new blood and skilled labor into areas with aging populations. But on the minus side, rural settlement slows the adaptation and integration of migrants. Resettlers, fearful of losing their employment-related housing, may be unable to quit exhausting and unappealing jobs. Tensions between locals and migrants over competition for scarce jobs and benefits and over different lifestyles are worse than in cities. Not least, the availability of a labor pool of low-paid, distressed migrants shores up inefficient agrarian enterprises.

Policies that adversely affect migrants' choices, such as residential restrictions, need to be changed. The goal should be to direct inflows toward cities, where prospects for resettlers are better in the long term. Urban housing should be made available, perhaps through municipal leasing. Finally, the government and the mass media should endeavor to overcome local residents' and authorities' negative stereotypes about migrants and to foster a more realistic view of migrants and the benefits they can offer Russian society.

Sergei A. Panarin "The Rural Economy of the Tunka Valley in a Time of Transition and Crisis"

A description of the precollectivization herding economy in Siberia's Tunka Valley and the changes introduced under collectivization provides background for an analysis of the effect of the economic transition on the Buryat people of Tunka and, in particular, the village of Tory. Fieldwork and data analysis show that the valley's population has not grown appreciably since collectivization, livestock numbers have declined, and the traditional mix of animals has been disrupted.

Under Soviet rule, social services improved, and in the early 1980s, when some centrally imposed requirements and restrictions were loosened, the system reached optimal productivity. The emphasis on grain and on nonnative strains of animals, however, led to depletion of resources and to a need for constant supplies of fertilizer and other externally provided inputs. With the loss of central support for the collective and state farm system, both the large farms and the household plots that depend on them have fallen into a precarious position. In particular, natural pasturage and fodder are inadequate to support the higher productivity of improved cattle breeds.

Rigid insistence on thoroughgoing market reform is not recommended under current circumstances. Private farms have failed to flourish so far, and household farms remain oriented toward subsistence. If the large-scale agricultural enterprises fail, the households that depend on them may be thrust into poverty.

Impact of the Transition: Approaches and Findings

L. Alexander Norsworthy and Olga Paluba

The rural reform in Russia has faced several hurdles, including a patchwork of regional and provincial experiments, and a lack of communications strategy and participatory capacity in the implementing institutions.

The results of development interventions have been mixed, not only because of the inherent difficulty of implementing uniform reforms over so vast a territory but also because of the differing strategies of the rapidly changing regimes in Russia

In order to develop a clearer picture of the impact of the transition in the Russian Federation's rural communities, several Russian scholars were invited to submit papers on a variety of rural and social development themes. A wide range of contributions was received, varying from national to local in scope and using rural economic, sociological, political, and demographic analysis to reach their conclusions. Some of the observations were consistent with those of World Bank experts in the same fields; others diverged widely from Bank findings in their interpretations of policy outcomes and institutions and in their policy prescriptions.

With the assistance of the rural and social development specialists in the Environmentally and Socially Sustainable Development Unit of the World Bank's Europe and Central Asia Region, these papers were reviewed, compared with other scholarship on the same subject, and edited for inclusion in this volume. The goal was to initiate a dialogue and an evolving research agenda in Russian institutions and in the Bank itself and so assist in the definition of research programs and project work able to address the pressing issues facing rural communities in Russia.

Approaches to the Analysis of Rural Russia

Agricultural economics as a discipline means substantially different things to a Russian today than in 1992. The intervening years have witnessed a transformation of the old agrarian structure of state and collective farms to a mixed sector made up of the remaining state-owned enterprises, joint-stock companies, partnerships, subsidiary farms, independent farms, and household plots. The process has been anything but smooth; a succession of fits and starts has been the dominant pattern.

The social sciences, including sociology, political science, and anthropology, as they are practiced in U.S. and European universities, draw on broader bodies of knowledge than did their Soviet cognate disciplines. Western scholars in these disciplines, however, are limited by unfamiliarity with the milieu and by lack of reliable, timely, and comparable data for Russia in the 1990s. Fieldwork is being done, but a theoretical framework is essential for assimilating and process the at times conflicting reports and observations.

Area studies, in general, have lost ground in recent years because of the dramatic cutbacks in university funding as a result of fiscal reallocations following the end of the Cold War. Nevertheless, many Sovietologists in the West have converted their paradigms to the new realities. Political scientists have exchanged the popular approaches of Kremlinology and institutional pluralism for new catchwords and phrases—"defense conversion" and "marketization," as well as privatization. Work on the subdiscipline of political science and international relations known as international development continues in Central and Eastern Europe and in the new independent states. That subdiscipline is able to accommodate the differing theoretical approaches of Russian, U.S., and European scholars to analysis of the evolution of the Russian rural sector.

During the Cold War, it was inappropriate to assume that democracy and freedom meant the same thing to Soviet scholars and policymakers as to their Western counterparts. The same is true of the concept of economic rationality today. To study rural Russia without a grounding in historical precedents is dangerous ground indeed, especially when for the Russians themselves, the implicit disciplinary bias is dialectical materialism and market economics is still a relatively new phenomenon.

The exact content of the discipline of economics—once the concern of managers, politicians, and various elites rather than of the general population—is now highly politicized in Russia. Land reform and enterprise restructuring have affected the grassroots. As the Russian elections in late 1999 demonstrated, the variegated political landscape can shift unexpectedly, with formerly influential parties disappearing entirely and new ones emerging.

Economists (if they have not simply refocused their inquiries on other regions) have moved from reconstructing socialist statistics to puzzling over data collection problems, the measurement of poverty, the impact of the black market, and the influence of the nouveaux riches. For economists who were Soviet specialists before the breakup of the U.S.S.R., the duality of the socialist system and the market has been a recurring theme. Reform was cyclical, even before the 1917 revolution and Lenin's New Economic Policy. For even longer, rural communities have been reactionary and slow to change for historical, social, and cultural reasons. As much as things have changed on the surface in terms of laws, government structure, and cabinet members, the local balance of power in rural communities has shifted less significantly than one might have expected when Boris Yeltsin first came into office and the new Constitution was adopted.

There has been a great deal of frustration in applying standard templates for development interventions to Russian rural areas. Many obstacles have been political and so lie outside the mandates of multilateral development banks. The gaps between legislation and implementation reveal a triangle in which the interdependent relations between *oblast* (regional) governments, collective farms, and households have remained relatively unaltered in many regions, even though the legal status of these entities may have formally changed. Party bosses continue to distribute subsidies, and managers continue to dispense special privileges and access to inputs. As Maria Amelina recently noted, we need to understand how it is that socialism can continue to exist in one sector—agriculture.

This volume takes a holistic approach toward examining the transition period and considers the economic, political, and social factors that are important in addressing the needs of rural and exurban communities. A reliable and universally accepted mechanism that determines which groups are poor and targets assistance to reach them is lacking and needs to be constructed.

We begin by sketching the background on agricultural reform and rural development based on studies from inside and outside the World Bank before turning to the papers prepared by Russian scholars for this collection. This introductory essay first reviews different approaches to the study of the rural sector in Russia. The chapter by Maria Amelina that follows contains an elaboration of a structural model.

In the section on structures, chapters by Vasily Uzun and by and Renata Yanbykh provide an overview of the institutional context and the substance of the land and farm reforms. The section on policy outcomes includes an essay by Vladimir Bogdanovsky on the social outcomes of land reform; analysis of a survey of 500 key stakeholders by Eugenia Serova; and a discussion by Zemfira Kalugina on the general paradoxes of the reform process. Serova's poll of decisionmakers well illustrates the conservative bent and the skeptical views on reform of the majority of agrarian opinion leaders. The section on adapative responses contains chapters by Kalugina exploring the survival strategies of households and enterprises and by Serova on changes in economic behavior. Other chapters deal with traditional and modern agricultural patterns and adaptation in a Siberian setting; the effect of the recent economic changes on rural women's work, household, and leisure activities; resettlement of involuntary migrants in Russia; and demographic shifts and land use trends in Moscow's periurban and exurban areas.

Recent Research: Findings and Data Problems

Much of the literature published by World Bank experts has proceeded from the operating assumption that if farms are privatized and market forces are unleashed, most distortions will be greatly reduced, unprofitable enterprises will be restructured and will cease to exist, and the state will play an increasingly small role—withering away, as it were.

The experience in Russia has shown, however, that economic reforms could not proceed at breakneck speed in that country in the absence of the necessary capacity and participatory mechanisms. These structures have to be built in municipal and local governments and newly restructured enterprises, with the assistance of nongovernmental organizations (NGOs) and private voluntary organizations (PVOs) in rural communities. Furthermore, social structures need to be developed and behavior change must take place before a market economy is fully functioning.

Joseph Stiglitz, former chief economist at the World Bank, attributed the failure or reforms in transition economies to the prescriptions of many Western advisers and their formulaic approaches. Price liberalization, privatization of state and collective farms, and land reform do not create a properly functioning agricultural sector in the absence of attendant institutional developments at the municipal, regional, and national levels. An understanding of the historical, cultural and social context is also necessary.

In the Europe Central Asia region as a whole, and particularly in the first and second tier of applicants for accession to the European Union, such as Poland, Hungary, Bulgaria, and Romania, recent research suggests that much information remains uncaptured by measurement of either household expenditure or household income. The findings show that off-farm income has the greatest potential for raising a family's standard of living above the poverty line.

Russian scholars point to the conservatism of the agrarian leadership and the rural population. Generally, rural heads of household are older than those in the country as a whole, and owners of independent farms are older than employees of other rural enterprises.

It is evident that many of the social assets of the former system have eroded or have ceased functioning altogether. Child care, subsidized heating, health care, schools, roads, and telecommunications infrastructure have been curtailed or discontinued. Exact statistics on this depletion of social capital, however, are not readily available from Russian scholars or from their Western counterparts. Similarly, the stopgap survival methods found in the informal sector of the rural economy have not been sufficiently documented. Survey research is plagued by the reluctance of the general population to disclose information, given the presence of this large informal sector.

The Institutional Setting: Structures

Until the late 1980s both production and inputs were subsidized by the central government. "Trying to shield consumers from the costs of an inefficient food and agricultural system by passing the costs on to the budget was unsustainable, and the system collapsed in autumn 1991" (World Bank 1992: 138). Price subsidies "so high that they exceeded government expenditures for health and education" (World Bank 1992: 139) led to market distortions. The reforms of the 1990s have reconfigured the traditional framework of property distribution and delivery of agricultural products to the market.

The first milestones of the reform were three laws adopted by the Russian Supreme Soviet in late 1990. The Law on Peasant Farms (December 1990) established the right to existence of collective farms. The Law on Land distinguished between private and state properties and legalized ownership of land by individual citizens, with the stipulation that land could be sold only to district governments and not to individuals.

The next significant legal development was a presidential decree of December 29, 1991, that made possible the reorganization of state and collective farms into joint-stock companies. The decree required farms to decide on their form of ownership (private or collective-shared) and to register with the state accordingly. It can be argued that this reorganization was nominal; although farm ownership generally did change, the organizational culture remained the same. Subsequent presidential decrees in 1993 and 1996 allowed for the sale and purchase of land shares back to the collective or new joint stock company. According to Uzun (1997), less than 5 percent of farm members desired to sell their land shares.

Table 1. Farm Entities and Forms of Ownership

Type of Ownership	Farm Entity
Collective	**Collective farm** (kolkhoz) **State farm** (sovkhoz) - non-land assets owned by enterprise **Limited liability partnership** (LLP or tovarishchestvo) - land and asset shares of the founders are pooled, and some or all of the founders work on the farm **Joint-stock company** (aksionernoye obshchestvo) - similar to LLP, stock certificates issued to owners according to the value of their land and asset shares **Association of peasant farms** - legally should consist of registered individual farms; in practice, it is another forms of reregistering collective enterprises **Agricultural cooperative** or **collective enterprise** (kollektivnoye predpriyatiye) - new name for some state and collective farms after reregistration
Private	**Peasant/family/private farm** (krestyanskoye/fermerskoye khozyaistvo) - farm based on privately owned or leased land (land size can range from 30 hectares in Moscow to 350 hectares in Saratov oblast) **Subsidiary (household) plot/farm** - belongs to employees of collective farms and has been reregistered under individual ownership **Private garden, vegetable and dacha plot,** also **subsidiary (household) plot** - belongs to urban residents

Source: Agriculture in Liberalizing Economies: Changing Roles for Governments, The World Bank, 1995: 148-152.

Legal and legislative reforms alone do not necessarily help overcome social, economic, and political obstacles. Nevertheless, Western analysts have tended to treat the restructured legal framework as the dominant factor in the development of rural Russia. The major structural problem with this approach is the gap between the legislation itself and its implementation, which has not always been adequate for the efficacy of reforms. People and their adaptive responses, not merely laws, require the most attention of policymakers if the aim is to reduce poverty and improve living standards.

Rural Reform: Policy Outcomes

Eight years after the Russian reform began, it can be characterized as a series of successes and failures. Success depends on the development of sustainable collaborative mechanisms between the three strategic actors in the communities: oblast and local governments, the large farms, and the residents. The remaining collective and state farms assumed dominant positions in the new economy of the countryside, according to Serova. The results for independent farms were mixed, with many new entrants and subsequent exits from the market.

Even with the liberal legislation on land and private property in place, the empowerment of farmers is minimal. A survey of farm managers and private farmers by the All-Russian Center for Public Opinion (VTsIOM) in early 1994 indicated that 34 percent of private farmers and 47 percent of farm managers felt that the presidential decree of December 1991 had had no effect on economic activities (Wegren and Durgin 1997: 4). Rural capital and labor stocks have declined. Current financial, tax, and credit policies do not favor private farming. Most private "urban farmers"—the owners of household plots and users of collective garden plots—are "farmers" only nominally but benefit from exemptions from VAT and land taxes (Wegren and Durgin 1997: 8). Private farms have experienced difficulties in marketing, processing, and delivering agricultural products. The main reason is state ownership of the supporting infrastructure.

Many people have left rural areas to seek work in urban centers. New, sometimes transient, workers have taken their places in many instances. In less fortunate areas where natural resource endowments do not favor crop growing by large, commercial enterprises, veritable ghost towns have replaced the state and collective farms. Vitkovskaya notes, "Small towns that were dependent on a single enterprise that could not survive in the post-Soviet period are economically gutted and depressed."

Before the crisis, financial losses of large farms were offset by budget transfers and subsidies. Today, with the diversification of agricultural entities, only a negligible part of subsidies goes to support private farms, and real subsidies have plummeted as a result of high inflationary pressures. Under the devolution of responsibilities from state to oblast governments in 1993, many oblasts lacked sufficient funds to provide adequate support for agriculture. Because of money shortages and the need to pay back loans, the amount distributed was rarely equal to the budget allocation for agricultural subsidies. Wegren (1998) presents evidence of increased bankruptcies and a downturn in the creation of new farms after the government suspended credits to private farms in October 1993.

The wedge between official and market prices has been growing, and the difference has been captured by the unofficial sector that is mostly involved in wholesale distribution of agricultural products. As a result, farmers and community residents are worse off than they were under the Soviet regime.

The central government and top-down planning have failed to deliver sustainable development solutions to rural areas. There is now a need to respond to the social changes resulting from the transition and to reduce poverty levels, especially among marginalized groups.

Survival Strategies and Demographic Trends: Adaptive Responses

The relationship between the former state and collective farms (some of which are now joint-stock companies), households, and municipal governments is one of interdependence. There are power asymmetries wherein one group's relative influence is issue-specific and the results of coalition building and negotiations with other groups on that issue. For example, private farmers may support the farm managers' position on subsidies because the farmers need inputs from the large farms, not because they agree wholeheartedly with the managers' approach. Political support for the subsidies may be relatively less important for the private farmers and relatively more important for the farm managers. In return, the farm managers may support the private farmers on issues that are less important to the managers but of vital importance to the private farmers. Managers may support the oblast government's limits on private farm size, not because it particularly affects their corporate interests but because the cooperation of the government is essential to secure subsidies, favorable tax treatment, and other benefits is more important than the goodwill of the private farmers in this area. Because households serve a double function—providing labor to the large enterprises and producing complementary or units—their bargaining power is greater in many cases than that of the independent farms. The strongest large enterprises are those which (i) seek to retain labor through better wages and the proviionof social services and (ii) provide inputs to the household and collective plots.

In other words, these negotiations between groups do not necessarily result in full compliance with the letter or intent of the laws or regulations on land or enterprise reform. The policy outcomes reflect the realities of the the local political balance of influence and access in each oblast. It has been argued elsewhere that agricultural reforms proceeded more smoothly in those oblasts where less was at stake—where agriculture accounted for a smaller proportion of total economic activity than other sectors.(see Amelina)

Regional and Local Governments

Among the more important areas under the control of regional and local governments are:

1. Disbursement of subsidies from the federal government to enterprises
2. The decision to implement land reform
3. The determination of the size of land shares for the region
4. Distribution of land shares to eligible employees
5. Conversion of land shares to landownership
6. Development of wholesale markets
7. Tax collection
8. Provision of social services previously funded by the state and collective farms

These powers enlarge the influence of local administrations and create a dependence of both large and small farms on the oblast government.

Agricultural Enterprises

Two general categories of agrarian enterprise exist within the rural sector in Russia. These are referred to as Categories A, B, and C in this essay and in the Conclusion (see Figure 1). Category A comprises primarily large, commercially viable enterprises (some remnant state and collective farms, along with new joint-stock companies). Category B consists of smaller household plots and collective garden plots, which are numerous and account for over one half of total agricultural output in the country. However, they do not have common ownership structures and are not restricted to rural areas. The commonality is that they are household based. Category C consists of independent farms and includes private, joint (partnerships), and leasing arrangements. Categories B and C rely for their survival on inputs from Category A entities and on subsidies and other incentives from regional or local governments. Households may have members employed in Category A enterprises while owning a household plot, a subsidiary farm, or a plot in a collective garden. Category B entities are responsible for a surprisingly large amount of produce and livestock, indicating that their contribution to household income is beyond the subsistence level (only producing for household consumption)—although it has not been captured in official statistics or studied systematically. Dairy, livestock and vegetable production show a share of total product of Category B entities that is disproportionate to the amount of land they control. Who is feeding the Russian population? The answer is both the large enterprises and the plots, certainly not the private "independent" farms.

Land distribution did not change significantly following the passage of land legislation. It is clear that most farmers assigned their shares to the reorganized collective farms. The lack of spontaneous and full decollectivization has surprised most Western observers. According to a 1994 World Bank survey conducted in five oblasts, there were no departures from 30 percent of the large farmsand only 1 to 9 departures from nearly two thirds (Brooks and others 1996: 33).

The link to a collective farm after institutional liberalization remains strong because the farm is in the first place a social community or network and only secondarily a production unit. It provides social services, protection from the chaotic economic environment, and inputs for the production of household plots.

Table 2. Basic Indicators, by Farm Type 1998

		Number	*Average size (hectares)*	*Percentage of output*
Category A	Former state farms	27,000	5,600	49.9
Category B	Household plots	16,000,000	0.4	53
	Garden plots	19,600,000	0.087	Not available
Category C	Private farms	274,000	48	2.2

Source: Economic Research Service, U.S. Department of Agriculture

Figure 1. Category A, B, and C enterprises

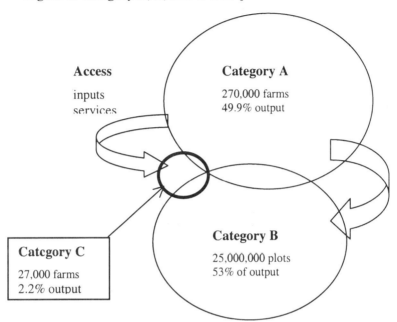

In general, would-be smallholders lack managerial experience and are reluctant to take risks. Most of the rural population has insufficient knowledge or understanding of formal rules and regulations. Moreover, potential independent farmers do not believe that they could survive without employment by and concessionary inputs from a large rural enterprise such as a Category A farm. Some authors, notably Uzun, suggest that the large Category A enterprises should function as service collectives for Category B enterprises. For this to work in each oblast and each municipality, the playing field between Categories A and B entities and the government would need to be leveled.

Households

According to the results of a poll by the U.S. Information Agency (USIA) released February 9, 1999, three out of every four Russians now grow some or all of their own food, a measure of the ways in which they are attempting to cope with their ever-increasing impoverishment.

Fifty-five percent of the population grows half or more of their food on private plots of land. Only 27 percent, the poll found, does not grow any of the food they consume —-in a country whose population remains more than 70 percent urban, where only 50 percent of Russian adults is employed, and only one in four of those who are employed is being paid on a more or less regular basis.

Fifty seven percent of those polled had borrowed money, and another 52 percent had accepted assistance of one kind or another from family or friends in the six months prior to the poll.

In Table 2, the percentages of total output indicate the overlap between the different types of entities. Garden plot output is primarily subsistence. Some of the household plots are being considered "subsidiary farms" and are included in the figures for the former state farms—just another indication of the blurred lines between the different types of entities.

Demographic Deficit and Migration

Erosion of the human capital base has been going on for many decades, but outmigration from rural areas has been especially evident during the transition period. Among the groups most affected by rural outmigration have been the skilled and the young. This has exacerbated the gap in economic development between cities and rural communities.

The weak population base was unable to produce a stratum of productive farmers in response to the reforms. In an effort to improve the demographic situation, the government has given urban dwellers incentives to relocate to the countryside. In spite of the benefits—payment of moving costs by the government, lump-sum compensation for those willing to move from urban areas, and assistance with housing, fuel, and social services—a significant influx of population into rural communities has not taken place.

Lessons from the Russian Research

Because institutions were not sufficiently developed, reform efforts were only partially successful. Land reform, in particular, was sporadic and varied greatly across regions. An important lesson from the Nizhny Novgorod pilot was that public communication of ownership rights, partnership with an NGO in organizing farmers' groups, and dissemination of public information were important ingredients. If reform is not to be subject to the whims of the oblast administration or the local Communist Party leader, strong institutions are needed to provide finance and business services and to maintain and expand the supply chain.

The leadership, as well as the public, has generally opposed reforms or has considered them ineffective because of the institutional gap. The Agrarian Party and its affiliated research institutes supported the land and rural enterprise reforms of the 1990s. They have, however, become disenchanted with and skeptical about central government initiatives and are calling for institutional transformations to shore up the reforms and reestablish momentum in rural communities.

The privately-owned farms and plots remain dependent on the remaining collective and state-owned enterprises or their successors, the joint-stock companies. The deterioration of less successful Category A farms deprives Category B farms of agricultural inputs needed for production, such as fodder, young animals, and loans of machinery. Economically hard-pressed Category A enterprises are also unable to maintain the social infrastructure (schools, health facilities, recreational centers, and the like), leading to additional hardships for their employees.

There has been considerable regional variation in the implementation of reforms. The degree of subsidies and other support for the agricultural sector is largely discretionary and remains with the oblast leadership. Rural communities vary in size, intensity of economic activity, and institutional structure, making generalization difficult. Each oblast has a particular distribution of negotiating power or clout, depending on the relative strength of rural enterprises and their proportion of the oblast's employment and income. The relative importance of agriculture then determines the interests of the stakeholders and the intensity with which they vie with other stakeholders for scarce municipal resources, including government subsidies.

The resource base and demographic trends are important considerations in evaluating behavioral responses to the reforms. In- and outmigration patterns affect labor markets and increase the pressure on the already depleted social infrastructure of rural communities. The resource base varies so widely among regions that the choice of product mix may be automatic for Category B farms. Other factors are proximity to urban centers, the market and distribution infrastructure of a particular region, and competing claims on land use. Keen competition for housing and real estate development has the potential to squeeze out the 2-hectare farms in favor of dachas for harried urban dwellers or large suburban residences for the nouveaux riches.

Toward a Political-Economic Model of Rural Russia

Economics is only one frame of reference for studying the behavioral and institutional responses to reforms. Valuable insights may be gained from sociology, economic geography, demography, and political science. In the scholarly and professional literature to date, each of these perspectives stands more or less alone. This volume underscores the need to combine the different approaches as a means of understanding the impact of reforms in Russia and finding new ways to smooth the effects of the transition. In an effort to integrate knowledge concerning the rural situation in Russia, Russian experts themselves need to be consulted and to participate in the process of developing policy recommendations.

Because of their formation under the Soviet system, Russian policymakers approach problems from a perspective that recognizes the importance of political as well as economic factors. Political economy is thus an appropriate approach for analyzing the outcome of reforms in the rural sector .The policy framework must be grounded in institutional and social analysis as much as in macro- or microeconomic modeling.

To develop effective economic development interventions, a carefully structured approach to the measurement of inputs, outputs, and impact should be applied. In this way, purely economic or legal factors will not be the only yardsticks for assessing the feasibility at the design phase, and the effectiveness during implementation of new policy initiatives. Behavioral change, institutional structures, and policy outcomes —in terms of public opinion regarding current problems, the success of initiatives, and their achievement of their intended economic and social effects —are proposed focal points for analysis. This requires a multidisciplinary approach involving political scientists and sociologists as well as agricultural economists.

Bibliography

Åslund, Anders. 1997. "Observations on the Development of Small Private Enterprises in Russia." *Post-Soviet Geography and Economics* 38 (April):191–205.

Åslund, Anders, ed. 1994. *Economic Transformation in Russia.* London: Pinter.

Braithwaite, Jeanine. 1995. "The Old and the New Poor in Russia: Trends in Poverty." Background paper for Russia Poverty Assessment. Education and Social Policy Working Paper. World Bank, Washington, D.C.

Brooks, Karen, and Zvi Lerman. 1994. *Land Reform and Farm Restructuring in Russia.* World Bank Discussion Paper 233. Washington, D.C.

Brooks, Karen, Elmira Krylatykh, Zvi Lerman, Aleksandr Petrikov, and Vasilii Uzun. 1996. *Agricultural Reform in Russia: A*

View from the Farm Level. World Bank Discussion Paper 327. Washington, D.C.

Prosterman, Roy L., Leonard J. Rolfes, Jr., and Robert G. Mitchell. 1995. "Russian Agrarian Reform: A Status Report from the Field." *Communist Economies and Economic Transformation* 7 (2): 175–93.

Lopez, Ramon, and Tim Thomas. 1999. "The Underlying Factors of Poverty in Rural Romania: Market Failure, Asset Productivity and Demographics." World Bank, Washington, D.C.

Serova. Eugenia. 1999. "Country Agricultural Policy Notes: Russia." In Alberto Valdés, ed., *Agricultural Support Policies in Transition Economies,* 29–48. Washington, D.C.: World Bank.

Uzun, Vasily. **1998**. *Reorganizing of Rural Enterprises: A Socioeconomic Analysis.* Moscow: Znak.

Vitkovskaya, Galina. 1999. "Adaptation and Integration of Forced Migrants in Russia." United Nations International Office on Migration, Geneva.

Wegren, Stephen K., ed. 1998. *Land Reform in the Former Soviet Union and Eastern Europe.* Routledge Studies of Societies in Transition 5. London: Routledge.

Wegren, Stephen K., and Frank A. Durgin. 1997. "The Political Economy of Private Farming in Russia." *Comparative Economic Studies* 39 (3–4, Fall-Winter): 1–24.

World Bank. 1992. *Food and Agricultural Policy Reforms in the Former USSR: An Agenda for the Transition.* Studies of Economies in Transformation 1. Washington, D.C.

World Bank. 1999. "Russia's Social Protection Malaise: How To Begin A Turnaround?" Human Development Sector Unit. Europe and Central Asia Region, Washington, D.C.

Rural Interactions in the Post-Soviet Era

Maria Amelina

At first sight, the idea of this collection of explicitly Russian views on the changes in rural Russia goes against the grain of post-Soviet developments. The end of the U.S.S.R. made it possible to break down the artificial barriers between different realms of Western and Soviet lives, including academic interactions. Russian scholars became a familiar sight on university campuses and in Western think tanks and research centers. Gradually, the work of Russian researchers, particularly on subjects related to post-Soviet restructuring, began to blend into the mainstream of Western academic discourse.[1] At the same time, for the first time in their careers, Western students of Russia had an opportunity to conduct hands-on research in the field and in archives.

Yet here is a volume dedicated to Russian contributions on the transformations in rural Russia. What makes "the Russian angle" different? What special insights can Russian scholars offer to Western audiences about the intricacies of Russian agricultural reform?

Western Expectations and Russian Reality

The Western vision of Russian rural reality and of Russia's agricultural reform was based on a number of economic and historical premises. The economic premise held that the Russian peasant, like any other rational economic actor, would choose an opportunity set that offered higher profit maximization. The spectacular and demonstrable inefficiency and unprofitability of collective agricultural enterprises led Western academics, and Western advisers to Russian reformers, to expect that Russian peasants would leave the shell of collective farming once the legal constraints that bound them to the collective were relaxed and individual farming, with its promise of higher returns, was legalized.

The economic argument for the unsustainability of collective farming was fortified by historical studies describing the brutality of Russian collectivization.[2] Studies by social scientists discussed the resentment

---

[1] As an example of a joint Western-Russian volume on a related subject, see O'Brien, Wegren, and Dershem (forthcoming). For a discussion of the efficiency of collective farms, see Sedik, Trueblood, and Arnade (1999).

[2] For studies of precollectivization and early collectivization policies, see Atkinson (1983); Conquest (1986); and Lewin (1968).

that peasants felt toward their collectivized predicament and the suffocating control of the state over agricultural inputs and outputs.[3]

This understanding of the low efficiency of collective production and the forced nature of peasants' association with the collective system was reflected in the recommendations made by Western scholars to Russian reformers. According to this advice, the role of the state was to create an adequate legal framework for individual farming. Once such a framework was created, the expectation was that the flow of peasants out of collective and state farms (*kolkhozy* and *sovkhozy*) would create a new reality of individual farming (see, for example, Prosterman and Hanstad 1995).

The reaction of Russian peasants belied these expectations: a decade into the reform, the number of collective farms has not changed. These enterprises remain the dominant agricultural producers, while the role of individual farms is still marginal and stagnant (Goskomstat 1997: 38; 1998).

Since agricultural reforms have not moved in the direction intended by the reformers and advisers, a number of hypotheses were put forward to explain the discrepancy between expectations and reality. The *legal* approach held that "farmerization" stagnated because of the incompleteness of the legal framework created for individual farming.[4] The *latifundista* explanation suggested that collective farm managements were not providing full information to employees about their right to exit the collective farm with land and property shares because rural elites hoped to appropriate land shares and concentrate agricultural land in their own hands (see, for example, Van Atta 1994). However, the land-titling campaign, aimed at improving the flow of information to the peasants about their rights to pursue individual farming, did not precipitate exit from collective enterprises.[5] And—an important point—although the legal framework indeed remained incomplete, there was no observable demand for a better one coming from the countryside.

In light of this low demand for the reform, students of Russian agriculture turned from efforts to understand the post-Soviet Russian environment to a closer scrutiny of the nature of the economic actors them-

---

[3] See, for example, Fitzpatrick (1994); Scott (1998).
[4] For a study that underscores the priority position of the legal reform, see Brooks and Lerman (1994). For similar views on the role of the legal framework in the success of individual farming, see Prosterman and Hanstad (1995); Van Atta (1993).
[5] A practical reflection of this effort is the World Bank Land Reform Implementation Support (LARIS) project, dedicated to the creation of land cadastres "as an information basis for national land policy" (World Bank, Staff Appraisal Report, May 16, 1994).

selves. It was argued that the Russian rural population was not embracing reforms because of its demographic characteristics—that old and aging rural populations were incapable of active participation in reform.[6] As I argue elsewhere (Amelina forthcoming, c), the demographic explanation has pitfalls: first, the connection between the age structure of the rural population and the propensity to reform is not clear, and, second, the explanation undermines the economic premise that rational economic actors choose economically more profitable solutions.

It is at this point that the insider perspective of Russian scholars can shed light on developments within Russian agricultural reality. As this volume demonstrates, through this perspective, Russian scholars can help Russian rural studies reconcile the puzzling behavior of Russian farmers with the expectations of a rational search for more profitable outcomes. The Russian angle may be described as a political-economic approach, and the papers presented in this volume deal with the economic and political aspects of the reform simultaneously. Considering the increased exposure of Russian scholars to the work of their Western colleagues, the political-economic approach is not simply an outcome of the special characteristics of Russian education in social sciences. As we try to show in this essay, it reflects post-Soviet institutional reality.

Costs and Benefits of the Collective Farm System: A Political-Economic Model

One of the subtle points that comes across from a close observation of transactions in the Russian countryside is the merged nature of the political and economic institutions of Russian agricultural production and agricultural governance. This unity is path dependent in nature, since all Soviet institutions, both economic and political, were created to serve political and economic goals simultaneously. The role of collective farms was not "just" to produce agricultural output but also to be an ideological showcase for the superiority of large-scale collective agriculture. Local governments did not exist "just" to provide a supportive environment for the operation of economic entities but also to coordinate the distribution of agricultural inputs and outputs and to oversee the timely fulfillment of plans.[7]

---

[6] See, for example, Ioffe and Nefedova (1997); Lerman (1999); Wegren (1998); and Wegren and Durgin (1997).
[7] On the interchangeable skill sets and roles of Soviet administrators and industrial managers, see Hugh (1969). For an analysis of "horizontal mobility between positions of political and expert authority" in Soviet agriculture, see Kaplan (1987).

The merged responsibilities of local governments and collective farm managers created an institutional alliance between these groups of actors (see Figure 1). The goal of their informal association was to guarantee preferential access to scarce inputs to the collective farm managers within the jurisdiction of a particular administrator. The professional success and material well-being of administrators depended on their ability to secure access to scarce resources for institutional clients.

Figure 1. A Post-Soviet, Access-Based Model of Interactions between Collective Enterprises and Local Administrations

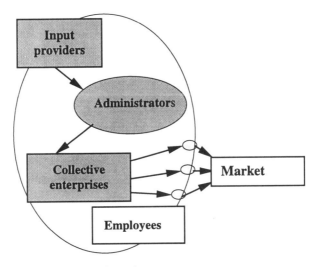

Source: Amelina (forthcoming, a).

The Soviet access-based system differs from a market-oriented ownership-based system in two important ways. First, within the access-based system, power and the flow of benefits are not directly connected to the ability to appropriate revenue on the basis of ownership; they depend instead on the ability to distribute resources efficiently within the network of associates. Second, the boundaries of the firm do not coincide with the legal definitions of these boundaries. Rather, the Soviet system was a *hierarchical, firm-like structure* that included collective farms and their employees, as well as local and national governments.[8]

---

[8] For more detail on the access-based model of post-Soviet distribution in agriculture, see Amelina (forthcoming, a)

Now, a decade into the restructuring, access-based distribution of resources has been officially abandoned by the federal government. Under the new design, the plan and the ability to secure access to resources are displaced by market prices, which are to determine relative scarcity. In this context the staying power of the predominantly unprofitable collective agricultural producers requires a more detailed investigation of the role that a path-dependent, access-based dynamics may play in the preservation of the traditional system. To understand the mechanisms that sustain this dynamics, it is important to examine the institutional structures that make the system operational in the new environment. To explain the motives that political and economic actors have for preserving the system, one needs to unbundle the political and economic benefits that accrue to the interested parties.

It is indicative that the papers in this volume do not attempt to study the economic restructuring of collective farms separately from the interactions between agricultural producers and the government (see the chapters by Serova and Uzun) and between collective farm employees and their employers (Bogdanovsky and Kalugina). How do these connections work in the post-Soviet context?

In a situation of weakly implemented federal agricultural policies, decisions about support for agricultural restructuring within a specific *oblast* (region) lie with the oblast government (Amelina forthcoming, b). The oblast government may choose to sustain the Soviet-style dependency of collective farms, for a number of reasons. One set of motives may have to do with the local administration's incentives to preserve the relevant relational capital and the skills mix. The change in the paradigm from an access-based to a market-oriented one undermines the accumulated stock of relational capital, as well as the value of the skills needed to successfully coordinate access-based interactions. A related set is professional: the access-based system allows an administrator to provide foodstuffs for budgetary institutions without drastically restructuring the provision of social services. It also "saves" the administration from having to restructure the provision of social services in the countryside and allows the collective farms to maintain their role as social protectors of last resort.[9]

Thus, partial revival of the old exchange practices allows local governments to achieve the double goal of supplying institutions with food-

---

[9] For a discussion of the difficulties local governments are likely to face in assuming responsibility for the provision of social services from industrial enterprises, see Alm (1995).

stuffs and minimizing the threat of strife and unrest in the countryside. Both are achieved by encouraging informal, path-dependent reciprocal contracts.

Another reason why administrators may want to stay involved in the distribution exercise in the post-Soviet era is the opportunity to seek rents from sales of remonopolized and accumulated commodities at higher prices outside the oblast. This motive, however, is not specifically post-Soviet; it is the same for authoritarian governments in developing countries or for any rent-seeking officials. In this sense, the interactions of Russian provincial governments with rural elites in general and collective farm management in particular can be analyzed within the frameworks used in the analysis of interactions between governments and interest groups, particularly in developing countries.

The incentives that collective farm managers have for continuing to depend on local administrators is analogous to those of local administrators: participation in the traditional exchange chain allows managers to continue with production and managerial practices without changing the skills mix, keeping accumulated skills and associations current and preserving them from being devalued by the new market paradigm. Within the post-Soviet contract, agricultural producers are responsible for producing agricultural output sufficient both to supply budgetary institutions and to provide for the needs of their employees. The enterprise does not provide for the needs of employees by paying a cash wage equal to the marginal cost of labor; to do that, managers would have to be able to sell agricultural output in a competitive market environment at prevailing market prices so that they could pay the wages. Support to employees comes, rather, in the form of in-kind payments that can then be used as inputs for individual production by employees on their private plots. Clearly, such goals are far removed from that of profit maximization, which is considered the primary objective for a market-oriented firm.

The ability of the employees-turned-shareholders to rely on collective farms for the provision of their basic needs, and the benefits that they derive from staying in an access-based structure, mainly depend on the discount they receive on the inputs (in-kind payments) obtained from the collective farm primarily for use in private production. This association can be termed *indirect profit maximization,* since the employees' strategy is not directly related to profit maximization by the collective enterprise as a viable firm but, rather, to their preferential access to resources preserved by the enterprise within the system. This chain of associations indicates that the relationship between the collective farm and its employees has evolved beyond the resentment and resistance of the postcollectivization period into that of mutually beneficial, if inefficient, cooperation.

The close interaction between political and economic exchanges has to be supported institutionally by a mechanism that, because of the low revenues of local governments, must be able to sustain access-based distribution without significant budgetary outlays. Such a system, termed commodity crediting, has been operational in most oblasts in Russia. Within the scheme, input providers receive tax forgiveness from local governments, valued at the amount of inputs granted to agricultural producers.

The ability to preserve access-based distribution without Soviet-style national coordination is based on two peculiarities of agricultural production. First, coordination does not have to be continuous. In the agricultural production cycle, inputs are required only twice a year, at sowing and harvest. Second, the low complexity of inputs (with Soviet-era equipment still operational, the required resources consist of fuel, fertilizer, and parts) means that local governments can preside over post-Soviet oblast-level distribution without national coordination.

If the local administration chooses to get involved in commodity crediting, it can control the "hardness" of the credit constraint for a particular season. If the political imperative of the oblast government is such that the support of the rural population is judged more important than the revenue that could be collected from the repayment of a commodity credit, the credit may be forgiven totally or partially. If increased revenues for the local budget take priority, local governments will make an effort to collect on the commodity credits. The budget constraint of an enterprise within this post-Soviet distribution structure can be termed elastic—harder when the repayment is enforced, and softer when it is not.

This peculiar post-Soviet arrangement does not take collective enterprises to the new level of market-oriented production with a well-defined priority of profit maximization. It does create a post-Soviet network of relationships based on profit plus maximization of political power. Within these relationships, employees' ability to maximize profits is constrained by the egalitarian nature of the distribution of resources, while the collective enterprises' impetus to restructure is constrained by the interventionist policies of local governments.

Of course, there is significant regional variance in this model. In some regions, local governments find continuous support to the system too costly and difficult to coordinate. Our research has shown that these are primarily regions in which agriculture is not a dominant sector and local governments do not have a professional history of agricultural management (Amelina forthcoming, b). Such a hands-off approach may ultimately lead to the demise of the unprofitable collective farms and to true restructuring of the more viable ones. However, national statistics on the resilience of unprofitable collective farms indicates the prevalence of the scenario that allows former kolkhozy and sovkhozy to survive is valid.

The papers collected in this volume present these multifaceted, hybrid political and economic interactions from different angles. The authors recognize, implicitly or explicitly, that the reform has not achieved separation of the political from the economic, and they explain the observed reality without the reference to expected outcomes found in many writings by more removed observers. The more involved position allows the reader to put aside the search for textbook expectations for the path of transition, which is supposed to foster viable capitalist farms, and to better tune into the actual dynamics that may lead to a new reality of political and economic rent-seeking.

One can hope that more focused advice on institutional change will be facilitated by the insights shared by the observers from within. Better understanding is a necessary first step for a more accurate reassessment of the potential for change in the Russian countryside and for better charting of routes that may lead to greater transparency and viability.

References

Alm, James. 1995. "Social Services and the Fiscal Burden in Russia." *Comparative Economic Studies* 37 (4): 19–31.

Amelina, Maria. Forthcoming (a). *False Transformations: From Stalin's Peasants to Yeltsin's Collective Farmers*

———.Forthcoming (b). "What Turns a Kolkhoz into a Firm?" In David O'Brien, Stephen Wegren, and Larry Dershem, eds., *Adaptation and Change in Rural Russia*. Washington, D.C.: Kennan Institute for Advanced Russian Studies.

———.Forthcoming (c). "Why Is the Russian Farmer a Kolkhoznik Still?" *Post-Soviet Geography and Economics*.

Atkinson, Dorothy. 1983. *End of the Russian Land Commune, 1905–1930*. Stanford, Calif.: Stanford University Press.

Brooks, Karen, and Zvi Lerman. 1994. *Land Reform and Farm Restructuring in Russia*. World Bank Discussion Paper 233. Washington, D.C.

Conquest, Robert. 1986. *The Harvest of Sorrow: Soviet Collectivization and the Terror-Famine*. New York: Oxford University Press.

Fitzpatrick, Sheila. 1994. *Stalin's Peasants: Resistance and Survival in the Russian Village after Collectivization*. New York: Oxford University Press.

Goskomstat (State Statistical Commission). 1997. *Statistical Bulletin 8* (October). Moscow.

———. 1998. *Sel'skoe Khoziaistvo v Rossii*. Moscow.
Hugh, Jerry F. 1969. *The Soviet Prefects*. Cambridge, Mass.: Harvard University Press.
Ioffe, Gregory, and Tatyana Nefedova. 1997. *Continuity and Change in Rural Russia*. Boulder, Colo.: Westview Press.
Kaplan, Cynthia S. 1987. *The Party and Agricultural Crisis Management in the USSR*. Ithaca, N.Y.: Cornell University Press.
Lerman, Zvi. 1999. "The Impact of Land Reform on the Rural Population in Russia." Paper presented at workshop on "Rural Russia," Kennan Institute for Advanced Russian Studies, Washington, D.C., May.
Lewin, Moche. 1968. *Russian Peasants and the Soviet Power: A Study of Collectivization*. Evanston, Ill.: Northwestern University Press.
O'Brien, David, Stephen Wegren, and Larry Dershem, eds. Forthcoming. *Adaptation and Change in Rural Russia*. Washington, D.C.: Kennan Institute for Advanced Russian Studies.
Prosterman, Roy L., and Tim Hanstad. 1995. *Land Reform: Neglected, Yet Essential*. Seattle, Wash.: Rural Development Institute.
Scott, James. 1998. *Seeing Like a State: How Certain Schemes to Improve the Human Condition Have Failed*. New Haven, Conn.: Yale University Press.
Sedik, David J., Michael Trueblood, and Carlos Arnade. 1999. "Corporate Farm Performance in Russia 1991–1995: An Efficiency Analysis." *Journal of Comparative Economics* 27: 514–33.
Van Atta, Donald, ed. 1993. *The "Farmer Threat": The Political Economy of Agrarian Reform in Post-Soviet Russia*. Boulder, Colo.: Westview Press.
Van Atta, Don. 1994. "Agrarian Reform in Post-Soviet Russia." *Post-Soviet Affairs* 10 (2): 159–90.
Wegren, Stephen. 1998. "The Conduct and Impact of Land Reform in Russia." In Stephen Wegren, ed., *Land Reform in the Former Soviet Union and Eastern Europe*. Routledge Studies of Societies in Transition 5. New York: Routledge.
Wegren, Stephen, and Frank Durgin. 1997. "The Political Economy of Private Farming in Russia." *Comparative Economic Studies* 39 (3–4, fall–winter): 1–24.

Agrarian Reform in Russia in the 1990s: Objectives, Mechanisms, and Problems

Vasily Uzun

During the socialist epoch Russian agrarian policy was based on the principle of state ownership. This policy led to an increase of the share of the state sector in agriculture (*sovkhozy* and other state-owned farms) and to a decline in the shares of the private (household plot) and cooperative (*kolkhoz*) sectors. By 1990 the private sector was using only 2 percent of agricultural land. Farmers could work their plots only in their spare time, when they were free from employment on the collective or state farm.

Although kolkhozy were called cooperatives, in fact they too were state-owned enterprises, and in the last decades of the system, the functioning mechanisms of collective and state enterprises were not much different. On both kinds of farm, agricultural production was carried out on state-owned land with state-owned means of production. (Kolkhoz capital assets were indivisible, and their purchase and sale were restricted, making them in fact like the capital assets of sovkhozy.)

Production was carried on according to state directives, and the products were mainly sold to government procurement agents at fixed government prices. Expansion of production was implemented according to the government's capital investment plan, mostly using government funds or state incentive credits (at 1–2 percent annual interest), subject to writeoffs over time. Workers in agricultural enterprises were paid according to state rates and norms of productivity.

Despite Russia's rich national resources and the high percentage of the population working in agriculture, the country's agrarian policy put it at a disadvantage in terms of crop, livestock, and labor productivity, compared with industrial countries. The principal reasons for the low efficiency of agricultural production were the deprivation of peasants' ownership of land and other means of production and the rigid centralization of agricultural management.

The need for radical changes in Russian agriculture had long become pressing. Efforts to improve the agricultural production system within the framework of the socialist economy invariably failed. The campaigns in the early 1960s to introduce intrafarm payments and collective and leasing contracts did not raise efficiency on collective and state farms. Efforts to make hired workers responsible for production costs and results came to nothing, nor was partial decentralization of planning successful. In 1986 the government allowed farms to sell at free-market prices 30 percent of planned volumes and all production in excess of the plan. This decision was either blocked locally or led to speculation and corruption, as market prices were much higher than state prices.

By the beginning of the 1990s it had become evident that if the problems in the agricultural sector were to be solved, it was necessary (in line with reforms in other sectors) to permit private ownership of land and other means of production and to provide opportunities for private methods of production. Transfer of ownership and a corresponding transfer of responsibility to individuals were considered necessary conditions for introducing changes in the management of agriculture and for improving efficiency.

Goals and Methods

The objective of Russia's agrarian reform of the 1990s was to foster efficient, competitive agricultural production based on private ownership of land and other means of production and on the initiative and entrepreneurial spirit of agricultural producers, with support from the state. A primary prerequisite for achievement of this objective was land privatization and farm reorganization. Many questions arose concerning the implementation of the reform.

Should landownership be transferred to individuals or to legal entities? Differences of opinion on this subject have not been resolved to this day. Clause 5 of the 1990 Law on Land Reform provided for transfer of land from kolkhozy, sovkhozy, and other agricultural enterprises to "collective [joint-share or equity-share] ownership." Later presidential decrees and government regulations provided that the land previously used by a given farm was to be transferred to the residents who were working or had been working on the farm. These workers were granted land share entitlements that they could use as they wished: they could start an independent (family) farm or expand the household plot, or they could sell, give away, exchange, or bequeath the shares.

The State Duma (the lower house of the Russian Parliament), in its drafts of the Land Code of Russia, interpreted clause 5 differently: state land being used by an agricultural enterprise was to be transferred to the ownership of that enterprise as a legal entity, and workers and pensioners of the enterprise could demand allocation of a parcel of land in exchange for a land share in order to create a family farm or expand a household plot. Under this approach, in 1992–93 agricultural enterprises were given the land by state acts and subsequently received certificates of landownership. Many managers of enterprises assume that these documents give an agricultural enterprise the right to use the land as it chooses.

To which categories of residents should land be transferred? Several options were discussed:

1. Return the land to those who had owned it before collectivization. Most Eastern European countries and the Baltic states took this option. Russia did not, for technical reasons: 70 years had passed since land nationalization, and it seemed next to impossible to find the old land entitlements. In addition, this course was politically unacceptable. In 1990 the Supreme Council of the Russian Federation rejected it, and clause 4 of the Law on Land Reform specified that land parcels could not be returned to the former owners, although these individuals could apply for ownership of land parcels on the same basis as other people.

2. Divide the land among the families living in the village at the time the reform took effect, in proportion to the number of family members or workers.[1] Albania, Armenia, China, and Vietnam chose this option. Russia rejected it, as it implied the dissolution of existing agricultural enterprises. The majority of rural residents were not ready for such a step, and there was a danger that the dissolution of kolkhozy and sovkhozy would destroy the infrastructure for servicing agricultural production (shops, depots, power stations, granaries, and so on) that had been created over many decades.

3. Divide the land into conditional land shares, distribute them among the workers and pensioners of an enterprise, and give each owner the right to make an independent decision on how to use the land share.

Russia chose the third alternative. Any individual can obtain a land parcel in exchange for the land entitlement and can use it as desired— for example, to create a family farm or expand a household plot. Land share owners can lease the shares to their agricultural enterprise, to farmers, or to other individuals. In this case the land parcels owned by lessors of the same enterprise remain within the collective and are not subject to physical division.

---

[1] A prominent Russian economist, G. X. Popov, proposed the following method: evaluate all privatized land, divide the total by the number of Russian citizens, and issue a land voucher to each person. It was assumed that agricultural enterprises and family farms would buy out or otherwise obtain vouchers from city and rural residents and so acquire land. This option did not gain the support of the legislators.

Other members of the Commonwealth of Independent States (CIS), except for Armenia, have chosen a similar formula. The advantages are that existing farms can be fully or partially preserved and land shares can be combined without preliminary physical separation. This approach avoids fragmentation of farms and obviates the need to divide Russia into 5–10 hectare (ha) plots as a preliminary to privatization. There are drawbacks, however: land share owners do not know where their land shares are located; managers of agricultural enterprises have an opportunity to use the land owned by citizens freely and without controls; and workers still, after nine years of reforms, do not clearly understand their choices. Most workers have not legally formalized transfer of land for the use of an agricultural enterprise and are deriving no economic advantages from landownership.

Should land be transferred to private ownership free, or for a charge? One opinion was that the transfers should be free because Russian peasants had already bought almost all the land under the 1861 reform that abolished serfdom and the Stolypin reform in the early 20th century. Others maintained that private owners should have to buy out state land because they would not use the land rationally if they obtained it free of charge.

Following extensive discussions and disputes in the mass media and in parliamentary commissions, a compromise was struck that provided for both kinds of transfer. Norms for free transfer of land were designated for each category of land user. In the course of privatization of kolkhoz and sovkhoz land, every agrarian worker, pensioner, and worker in the farm's social infrastructure had a right to a free land share. The size of the share was determined by dividing the total land area of the enterprise by the number of applicants for land shares. The size of the share, however, could not exceed the mean size of a land share in the given administrative area.[2] Local government bodies fixed the appraised quality value of the land share in point-hectares and determined and approved the size. If the size of a land share (taking into account the quality of the land) in a given enterprise was less than the average norm in the *rayon* (district), all the agricultural land of the enterprise was privatized free of charge. If a farm had more agricultural land than the average district norm for workers and pensioners, these applicants obtained a land share equal to the norm; any remaining land stayed in state ownership and was transferred to the special district fund for redistribu-

---

[2] The average land share for Russia is 10 hectares. The size of land shares in different administrative areas of Russia varies from 2 hectares in Chuvashia to 22 hectares in Volgograd, Orenburg, and Novosibirsk *oblasts* (regions) and 34 hectares in Chita oblast. (Data, from Goskomzem 1999, are as of January 1, 1999.)

tion. Citizens who had not obtained land shares through privatization of kolkhoz and sovkhoz land could obtain a land parcel for a family farm free of charge from the special district fund, but the size of the parcel could not exceed the district norm multiplied by the number of workers on the farm. Extra land could be obtained for a charge.

In addition, subnational government bodies within the Russian Federation define the maximum size of agricultural land owned by a family farm, irrespective of how the land is acquired. The size depends on the natural and climatic conditions in the region and varies between 30 ha and 200 ha. A family farm can lease unlimited amounts of land above this norm.

Privatization of Different Forms of Property

Household plots. Owners of household plots were allowed to privatize, free of charge, the land that they had used before the reform and to obtain an additional parcel free, but the total area of their land could not exceed the maximum size approved by local government bodies. Usually, the size of a household plot does not exceed 1 or 2 ha.

Garden and vegetable plots. Garden and vegetable plot users could become owners of the land they had been using or could obtain new plots in agricultural areas with poor soil or in nonagricultural areas. The size of a plot allotted to a family could not exceed the maximum size approved by local government bodies (usually, 600 to 1,200 square meters).

Agricultural enterprises. Privatization of the property of agricultural enterprises was carried out by dividing the cost of the property (less all the debts of the enterprise) by the number of property shares. Social infrastructure (schools, hospitals, clubs, and the like) was not included in the privatized property. The size of an individual property share depended on the length of service and the salary history of the worker or pensioner of the enterprise.

Each share owner was granted an opportunity to obtain a parcel of land equal to the cost of the share for use as a family farm; to demand monetary compensation for the share; to sell, give away, or bequeath the share; or to contribute it to the share capital of agricultural joint-stock companies, limited liability companies, or cooperatives that were founded in the process of reorganizing the collective and state farms. Exceptions to these rules were made only for large state livestock farms, greenhouses, and poultry farms that had been constructed with state capital investments. On these farms, individual shares were calculated by subtracting from the total value the state share (including the cost of productive assets) as measured by budget funding and centralized investments for the 15 years preceding privatization. The state share was transferred to the regional state property fund, which could sell it to family farms or to new agricultural enterprises created through reorganization, with payment to be made in installments over three years.

The Outcome

It is obvious that as a result of privatization, property (except that obtained by private farms in exchange for land entitlements) was transferred to agricultural enterprises, not individuals. Former members of kolkhozy and workers in sovkhozy became stakeholders (owners of shares, stocks, and so on), and enterprises as legal entities became property owners.

The Russian program of privatizing the land and property of agricultural enterprises was explicitly based on the principle of social justice. In allotting property free of charge, it was important not to harm the interests of the main social groups in rural areas. Accordingly, pensioners were granted land and property shares. Although it was clear that they would not be able to work on their land, they could give or bequeath their ownership to their children or grandchildren and could supplement their pension income by leasing property or by selling their land and property shares. On the same grounds, workers in social infrastructure located on the farm obtained land shares.

The basis of the ideology behind agrarian reform in Russia was the assumption that workers granted land and property entitlements would make use of their rights to leave the sovkhozy and kolkhozy, take their land and property, and start their own farms. It was assumed that as a result of privatization, agriculture in Russia would be mainly represented by family farms—that is, it would have the same structure as agriculture in industrial countries.

These expectations were not fulfilled. Only 5 percent of the workers left their enterprises and started their own farms. The rest did not take the risk of starting independent enterprises. Rather than make a quick transition to large, commodity-producing family farms, the majority of rural residents preferred two alternatives that seemed more understandable, simple, and traditional:

- *Expansion of household plots.* In the socialist era everything possible had been done to restrict this sector: the plots were cut off, livestock numbers were limited, and having one's own means of production (horses, minitractors, and the like) was prohibited. Work on household plots could be done only in the worker's spare time. With the end of the restrictions, household plots expanded and became the main source of income for rural families.

- *The formation of large enterprises by rural entrepreneurs* who used the land and property shares of rural residents and hired workers. Gradually, many former kolkhozy and sovkhozy became private enterprises managed by one or a few active entrepreneurs.

The Legislative Base for Privatization

Privatization of agricultural land and property in Russia was carried out on the basis of the 1990 amendments to the Constitution of the Russian Federation; the laws of the Russian Federation "On Land Reform" (1990) and "On Peasant Farms" (1990); the Land Code (1991); Presidential Decree 323, "On Urgent Measures for Implementation of Land Reform in the Russian Federation" (1991); and Regulations 86, "On the Procedure for Kolkhoz and Sovkhoz Reorganization" (1991), and 708, "On the Procedure for Privatization and Reorganization of Enterprises and Organizations of the Agroindustrial Sector" (1992).

The constitutional amendments of 1990 canceled the state monopoly on land; legalized a variety of ownership forms, including private ownership; and proclaimed that the property of kolkhozy and sovkhozy was owned by residents working on these farms.

"On Land Reform" established the procedure for transferring land (a) to the ownership of residents for a household plot or family farm or for gardening, livestock keeping, and other purposes related to agriculture, and (b) to collective (share or common) ownership by the members of kolkhozy and other agricultural enterprises. It also laid down the guidelines for transferring land free or at a charge. A 10-year moratorium on the purchase and sale of agricultural land was introduced; during that period land could be sold only to the state. (The 1990 law was canceled in December 1993 following adoption of the new Russian Constitution.)

"On Peasant Farms" set out the procedure for allocating municipal land, land parcels, and property to family farms in exchange for land and property entitlements.

The Land Code, prepared in compliance with the Law on Land Reform, contained a detailed description of the rules. It regulated the turnover and use of agricultural and other land and spelled out the procedure for determining the ultimate sizes of land parcels for different categories of users.

In spite of the constitutional amendments and the related laws, privatization and reorganization of agricultural land and property were not widespread in 1991. Neither de jure nor de facto transfer of land and property from state ownership to the ownership of members of agricultural enterprises occurred. One reason was the lack of mechanisms for privatization and reorganization and the bureaucratic complexity of the process. For instance, each enterprise had to obtain an order from the Russian Ministry of Agriculture to allocate land and property shares to its members.

To accelerate and streamline the process of land privatization and farm reorganization, Presidential Decree 323 and Regulation 86 were adopted in December 1991. These documents obliged state bodies and agricultural enterprises to legally formalize transfer of ownership of land and property from the state to members of agricultural enterprises and thus bring the legal forms of agricultural enterprises into compliance with the new ownership relationships. The decree and the regulation mandated completion of land privatization and farm reorganization during 1992.

Regulation 708 clarified some issues of land privatization and reorganization relating to kolkhozy and sovkhozy. It also elaborated on the conditions for privatization and reorganization of specialized farms, state livestock complexes, greenhouses, and poultry farms.

In spite of countermoves by the opposition, the majority of agricultural enterprises implemented privatization and reorganization in 1992; all others completed these activities in 1993. Such adherence to the law, rare in Russia, may be explained by the interest of farm managers and specialists in formally transferring land and property from the state to farm members—and, in reality, from the state to farms. In 1992–93 practically all agricultural enterprises obtained state acts that indicated the total area of transferred land, including land in collective (share and common) ownership. Lists of owners were attached to the acts, but that did not hinder farm administrations from using the land as if it were owned by the legal entity rather than by residents.

As a result of several circumstances—the short time allowed for privatization and reorganization, the almost total lack of information and communication activities among residents, and managers' fears that transfer of land and property to farm members would result in enterprises being split up—privatization and reorganization were implemented only formally. Owners received no documents for land and property and did not understand their rights and opportunities. Farm managers did not know how to start an agricultural enterprise under market conditions, were not aware of differences between the various legal forms, and lacked knowledge of how to build relationships between an agricultural enterprise and owners of land and property shares.

The Nizhni Novgorod model of land privatization and farm reorganization was developed to deal with all these problems. That model was invoked in Presidential Decrees 1767, "On Regulation of Land Relations and Development of Land Reform" (1993), and 337, "On Realization of the Constitutional Rights of Citizens to Land" (1996), and in Regulations 874, "On Reorganization of Agricultural Enterprises Regarding Experience of Nizhni Novgorod Oblast" (1994), and 95, "On the Order of Exercising Rights of Land and Property Share Owners" (1995). These instruments, it should be emphasized, were adopted not to regulate the process of redistributing ownership but rather to provide rural residents with an opportunity to make use of their rights as owners and derive economic advantages from ownership. They were also meant to assist managers of agricultural enterprises and farmers in correctly building up their relationships with land and property owners.

Stages in Implementation of the Reform

The course of agrarian reform in Russia may be divided into several stages:

1. Privatization, price liberalization, transition to the market, formal reorganization of agricultural enterprises, and the formation of a private farming sector

2. Realization of the constitutional rights of citizens to land and property, development of land leasing, reorganization of agricultural enterprises and harmonization of their charters with the Civil Code of the Russian Federation, formation of market infrastructure, and the establishment of a credit system in the agrarian sector

3. Financial recovery and formation of competitive enterprises, concentration of the capital of agricultural enterprises by rural entrepreneurs, expansion of the private farming sector in the form of household plots, formation of a state system to regulate and support the agroindustrial sector, and development of a land market.

The reform has been proceeding at different speeds across the country, and so it is difficult to define time spans for the separate reform stages. The first stage started in 1991 and was mostly completed in 1993, although not in all regions. The second stage began in 1994 and is not yet complete. Preparatory work for activities of the third stage is under way.

During the first stage, the state monopoly on land was liquidated. As of the beginning of 1998, 138 million ha of agricultural land (62.6 percent of the total area) had been privatized and transferred to citizens. The total included 116.2 million ha transferred to land share owners (11.8 million people), 11.6 million ha to family farms (270,000 farms), and 10.2 million ha to household plots (16 million families) and to garden, vegetable, and *dacha* plot owners (22 million families).[3]

Almost all agricultural enterprises went through reorganization and reregistration during the first stage of reform. About 10 percent of agricultural enterprises remained in the state sector; the rest reorganized into joint-stock companies, limited liability companies, and agricultural cooperatives. Some agricultural enterprises reregistered during the first stage of the reform and retained the form of a kolkhoz or sovkhoz, but the number of personnel in such farms decreased each succeeding year, and many reorganized into other legal forms. Some enterprises have not merely reorganized but have undertaken deep reforms, splitting into family farms or several smaller enterprises, creating associations of family farms, and so on.

An important result of the first stage of agrarian reform was the formation of a private farming sector. Some private farms, especially at the beginning of the reforms, were created by town residents who obtained land from special district land redistribution funds. In the course of mass reorganization of agricultural enterprises (1992–93), private farms were mostly created by former workers of kolkhozy and sovkhozy who did not transfer their ownership to enterprises created through reorganization but applied for physical allocation of land and property to start private farms.

The second stage of the reform, which started with development and implementation of the Nizhni Novgorod farm reorganization model, was aimed at giving land share owners a choice as to how to use their ownership; providing rural entrepreneurs with an opportunity to start new enterprises; and bringing the founding charters of existing farms into compliance with the Civil Code. Certificates of ownership of land shares were issued to the majority of owners (10.9 million people). As of January 1, 1999, 842,000 people (7.2 percent of the share owners) had received land parcels for their land shares, and about 5 million owners (42.5 percent) had agreed to convert their land shares to leases to agricultural enterprises and private farms. About 5 percent of owners

---

[3] Information on the progress of the land reform as of January 1, 1999, is from Goskomzem (1999); information on household plots is from the Russian Agroindustrial Committee.

concluded other types of agreement for using land shares—for example, by contributing to the share capital of their enterprises (Goskomzem 1999). The process of concluding agreements is continuing.

In the sector of individual plots and of garden, vegetable, and dacha plots, an active land market has grown up. In 1997 more than 400,000 land transactions were completed in this market.

The process of reorganizing agricultural enterprises is still under way. It became particularly active after the adoption of the Civil Code and the laws "On Joint-Stock Companies" (1995), "On Agricultural Cooperatives" (1995), and "On Limited Liability Companies" (1998), which stipulated, for each legal form, the terms and procedures for either bringing documents into compliance with the law or reorganizing.

In 1998 preparations for the third stage of reform began. These included development of models for financial normalization and reorganization of insolvent enterprises; development of household plots; expansion of the farming sector on the basis of household plots; establishment of mechanisms for concentrating agricultural enterprises' land and capital in the hands of effective owners; creation of systems of state regulation of and support to the agroindustrial sector; and development of land market infrastructure.

Changes in Agrarian Structure as a Result of the Reforms

Modern agriculture in Russia is represented by three sectors: large agricultural enterprises, household plots, and family farms.

Large agricultural enterprises. At the beginning of 1998 there were 27,000 enterprises, which employed 60 percent of all agricultural workers and held 69 percent of agricultural land. In 1997 their share of gross agricultural product was 49.9 percent. The average agricultural land area held by these enterprises was 5,526 ha, and the average number of workers was 212 (Goskomstat 1998: 451).

Although the agricultural enterprises in this group have different legal forms (joint-stock companies, limited liability companies, mixed partnerships, and kolkhozy and other forms of collective enterprise), all of them are, in fact, production cooperatives because of their large number of members (500 to 700 stockholders, share owners, or stock owners), each with one vote. Neither the managers nor the workers of the majority of these enterprises are sufficiently motivated to increase their efficiency. The evolution of these enterprises toward private companies is one of the main goals of the reform.

Household plots. There are 16 million household plots, with an average size of 0.67 ha. This sector accounts for approximately 35 percent of all labor resources in agriculture. In 1997 household plots produced 47.9 percent of gross agricultural product and accounted for 66.6 percent of gross value added in agriculture.[4] Their share of agricultural land is 4.8 percent, and they also use 16 million ha (7.2 percent) of the agricultural land of local rural administrations. The greatest part of the fodder produced by collectively owned enterprises is distributed among enterprise members and is used on household plots.

Development of household plots and transformation of some of them into commodity-producing private farms, along with transformation of collectively owned farms into cooperatives servicing household plots, are important future goals.

Family farms. There were 270,000 family farms at the beginning of 1999. They accounted for about 6 percent of agricultural land and 2 percent of gross agricultural product. The average family farm has 51 ha of agricultural land and 3 workers. The process of forming new family farms and expanding existing ones is continuing. In 1997–98 alone, the agricultural land area of household plots increased by 1.6 million ha. Opposite processes are going on simultaneously, as family farms cease operations and transfer their land back to the state or to collectively owned farms.

Thus, the main agricultural producers in Russia are, on the one hand, the largest enterprises, which, in land area and especially in number of workers, are tens and hundreds of times larger than private farms in industrial countries, and, on the other hand, the smallest household plots, with areas that are hundreds of times less than the size of private farms in industrial countries.

At the beginning of the agrarian reform in Russia, there were two forecasts for its development. Some assumed that, given the right to land, peasants would create millions of private farms and would become major agricultural producers. Others argued that after elimination of centralized planned orders and controlled prices, large agricultural enterprises would increase their efficiency under market conditions and drive smaller producers out of the agrarian market. Neither forecast came true. In fact, during the reform years, the importance of household plots in agricultural production increased rapidly, and the role of large enterprises declined. In 1990 the share of household plots in gross agricultural product was 26.3 percent; by 1997 it had grown to 47.9 percent (Table 1).

---

[4] Outputs and gross value (in current prices) for 1997 are from Goskomstat (1998): 3.

Table 1. Gross Agricultural Product by Type of Farm, 1990–97

(percentage of total)

	1990	1991	1992	1993	1994	1995	1996	1997
Agricultural enterprises	73.7	68.8	67.1	57.0	54.5	51.5	50.8	49.9
Household plots	26.3	31.2	31.8	39.9	43.8	46.6	47.4	47.9
Family farms	—	—	1.1	3.1	1.7	1.9	1.8	2.2
Total	100	100	100	100	100	100	100	100

— Not available.

Specialization of farms of different types became evident. Large enterprises retained grain and fodder crop production as their priority, while household plots concentrated on potatoes, vegetables, and fruits and berries (Table 2). The share of household plots in animal products is increasing (Table 3).

Table 2. Main Crops Produced by Different Types of Farm, 1991–97

(percentage of total production)

	1991	1992	1993	1994	1995	1996	1997
Grain							
Agricultural enterprises		97.4	94.2	94.2	94.4	94.6	93.0
Household plots		0.5	0.6	0.7	0.9	0.8	0.8
Family farms	0.2	2.1	5.2	5.1	4.7	4.6	6.2
Sugar beet							
Agricultural enterprises		97.8	95.8	95.8	95.9	96.0	95.7
Household plots		0.2	0.3	0.7	0.6	0.7	0.8
Family farms	—	2.0	3.9	3.5	3.5	3.3	3.5
Sunflowers							
Agricultural enterprises		93.0	88.6	88.2	86.2	87.0	87.8
Household plots		1.2	1.5	1.6	1.4	1.6	1.4
Family farms	0.4	5.8	9.9	10.2	12.4	11.4	10.8
Potatoes							
Agricultural enterprises	27.7	21.2	16.5	11.0	9.2	8.9	7.7
Household plots	72.0	78.0	82.5	88.1	89.9	90.2	91.3
Family farms	0.3	0.8	1.0	0.9	0.9	0.9	1.0
Vegetables							
Agricultural enterprises	53.8	44.5	34.5	32.2	25.3	22.1	22.2
Household plots	46.2	54.7	64.5	66.7	73.4	76.8	76.3
Family farms	—	0.8	1.0	1.1	1.3	1.1	1.5

— Not available.

Source: Russian Agroindustrial Complex, *Annual Statistics,* 1991–97.

Table 3. Animal Products by Type of Farm, 1990–97

(percentage of total production)

	1990	1991	1992	1993	1994	1995	1996	1997
Cattle and poultry for slaughter								
Agricultural enterprises	75.2	70.1	64.7	60.0	55.4	50.1	46.9	43.2
Household plots	24.8	29.7	34.5	39.5	43.2	48.3	51.4	55.0
Family farms	—	0.2	1.1	1.2	1.4	1.6	1.7	1.8
Milk (gross yield)								
Agricultural enterprises	76.2	73.9	68.1	64.2	60.0	57.1	53.1	51.4
Household plots	23.8	26.0	31.4	34.7	38.7	41.4	45.4	47.0
Family farms	—	0.1	0.52	1.1	1.3	1.5	1.5	1.6
Eggs								
Agricultural enterprises	78.4	77.8	73.8	72.7	70.9	69.4	68.4	69.3
Household plots	21.6	22.2	26.1	27.1	28.8	30.2	31.2	30.4
Family farms	—	—	0.1	0.2	0.3	0.4	0.4	0.3
Wool								
Agricultural enterprises	75.5	71.7	67.0	62.8	60.2	52.7	49.3	44.0
Household plots	24.5	28.2	32.2	35.4	37.1	42.8	46.1	51.3
Family farms	—	0.1	0.8	1.8	2.7	4.5	4.6	4.7

— Not available.

Collectively owned farms have reduced their production and their number of workers. Many people fired from or occupied only part-time in agricultural enterprises are working on household plots. Household plots adapted to market conditions sooner than did collectively owned farms.

Since production of household plots is not subject to taxation, their actual income per cost unit, particularly in large-scale production, is higher than that of collective farms. Moreover, the product of collectively owned farms is often not officially accounted for but is immediately transferred to household plots.

Factors Restricting Agrarian Reform

The reasons why agrarian reform has been uneven and incomplete are many and diverse. They include psychological factors and ideological differences, institutional shortcomings and inadequate resources, the larger economic situation and the narrow self-interest of individual actors.

Shortage of initiative and entrepreneurship among peasants. For decades, people who showed initiative left or were forced out of rural areas. Those who stayed got used to being hired workers and to having specific assignments such as milking cows and driving tractors or cars. In the reform period there was a need for rural entrepreneurs with initiative who were ready to start private farms or small agricultural enterprises and who knew how to operate machinery and equipment, organize production and accounting, and evaluate decisions. There are few such people in Russia's rural areas.

Price disparities. The agrarian reform is being implemented simultaneously with the general economic reform. Liberalization of prices led to disparities between agriculture and industry; the prices of production inputs rose approximately five times as rapidly as those of agricultural products. Agricultural production started to suffer losses or became unprofitable. Few people venture to start their own business in such conditions. It is hard to find people willing to participate in the reorganization of collective and state farms, as many farms have huge debts that would pass to their legal successors on reorganization.

Conflicts between the executive branch and the legislature. The executive branch champions the transfer of land to peasant ownership, purchase and sale of land, and mortgages and leasing. The legislative branch advocates transfer of land to agricultural enterprises, a moratorium on land transactions and on the mortgage of agricultural land, and restrictions on leasing land. This conflict has gone on for many years. It has resulted either in passage of contradictory laws, with both ideologies represented, or in blockage of laws. The clearest example is the Land Code, which has been discussed for seven years now. The standoff exacerbates the doubts of administrators at all levels and undermines executive discipline. The irrevocability of reform is still in doubt because decisions either are not made or are delayed for long periods.

Unclear division of competence between federal and regional power. Many regions do not implement federal legislation, referring instead to local decisions and to referenda in which the majority of the people voted against private ownership. Fifteen main administrative divisions of the Russian Federation (out of 89) still have not completed the activities of the first stage of agrarian reform. In particular, they have not transferred agricultural land from state ownership to ownership by workers. The lists of land share owners have not been completed, and land entitlements have not been issued. As a result, the rights to land of about 2 million–3 million rural residents have been ignored.

Lack of a federal body to manage agrarian reform. In 1990 a Russian Committee on Land Reform was created; it lasted for a year. Then a Centre of Agrarian Reform under the vice president, A. V. Rutskoy, was created, but it was dissolved before it began developing its activities. Since then, many organizations have been responsible for implementing agrarian reform in Russia, but none deals specifically or exclusively with this issue. (The Ministry of Agriculture has responsibility for reform along with thousands of other problems.)

Some jurisdictions, such as Orel and Tumen oblasts, created centers of agrarian reform, funded from local budgets. Agrarian reform is implemented more actively in such cases.

Low level of executive discipline. Officials often delay in carrying out decisions on implementation of reforms. For example, since 1991 several presidential decrees and governmental regulations have mandated issuance of entitlements to land share owners. As of the beginning of 1998, however, approximately 1 million people (8 percent) had not received entitlements. Many farm managers put the certificates in the safe instead of handing them out to owners.

Decree 337 (1996) ordered all managers of agricultural enterprises to conclude lease agreements or other agreements with land share owners. Nevertheless, at the beginning of 1999 agricultural enterprises had signed agreements with less than 45 percent of land share owners, although they are using the share owners' land. Under Russian law actions of this kind constitute unwarranted seizure of land subject to criminal prosecution, but the legislative norms are not applied in such situations.

Lack of funding. Reform implementation requires significant funding—for information campaigns, to develop mechanisms for implementing the reforms, and to train officials and specialists in methods of implementation. The federal budget, however, does not provide such funding. Many reform activities were developed and implemented with funding by sponsors—most often, foreign sponsors.

Conflicting interests. Managers and specialists of agricultural enterprises have a stake in preserving the enterprises intact. One motive is to secure their jobs. Another is to accumulate, over time, the capital and land in their own hands and then to become large rural entrepreneurs, transforming the collective farm into a private enterprise.

The attitude of common workers toward the fate of agricultural enterprises is ambivalent. Their individual plots are more important to them, but they fear losing their jobs within the agricultural enterprise. Those jobs ensure some income, tenure, guarantee of a pension allowance, and such benefits as paid sick leave and maternity and paternity allowances. In addition, work on the farm provides access to means of production (tractors and other agricultural machinery, automobiles, fuel, and so on) and to collective farm products, such as fodder, that are used in household plots, usually gratis or for a small fee. Workers are therefore interested in preserving the agricultural enterprise and transforming it into a cooperative that provides services to household plots.

The predominance of personal over collective interests has brought embezzlement and ruin to many collectively owned farms. Only their transformation into private companies or cooperatives that support household plots can stop that trend. The direction in which collectively owned farms evolve (into large private enterprises, service cooperatives, or a number of smaller enterprises) depends on many factors, and the process is likely to take place over an extended period.

References

Goskomstat (State Statistical Commission). 1998. *Annual Russian Statistics.* Moscow.

Goskomzem (State Land Commission of the Russian Federation). 1999. *Annual Russian Statistics.* Moscow.

Russian Agroindustrial Complex. Various years. *Annual Statistics.* Moscow.

Driving Forces in Russian Agrarian Policy in the 1990s*

Renata Yanbykh

During the Soviet era, the main and absolute driving force in agrarian policy was the Communist Party of the Soviet Union (CPSU). All decisions were made, implemented, and controlled by the CPSU through its republic, regional, and local branches. The reforms of the 1990s and the introduction of the new Russian Constitution in 1993 brought about radical changes in this structure. The bodies now participating in agrarian policy include the Federal Assembly, the president and the executive branch of government, political parties and blocs, social organizations, business and producers' associations, and research and consulting groups.

The Government

The Legislature

The Federal Assembly, established in 1993, consists of the State Duma (the lower house, with 450 members elected for four-year terms) and the Federation Council (the upper house, with 178 members, 2 from each constituent member of the Russian Federation). The major political parties compete for seats in the State Duma (Table 1).

According to Article 19 of the State Duma's regulations, the main tasks of the Duma's 28 committees are to:

- Review, with expert advice, drafts of legal acts

- Prepare resolutions on laws and amendments to laws

- Initiate parliamentary discussions of specific issues

- Comment on the relevant articles of the draft federal budget.

- Under Article 13 of the Federation Council's regulations,

---

* This report was prepared under the supervision of Eugenia Serova at the Analytical Centre for Agrifood Economy of the Institute for Economy in Transition.

the main tasks of the Council's 11 committees are to:

- Comment on draft federal laws and amendments to the Constitution approved and recommended by the State Duma
- Make proposals on legal acts of the Parliament of the Russia and Belarus Union and on model legal acts approved by the Interparliamentary Assembly of the members of the Commonwealth of Independent States (CIS) and the Interparliamentary Committee of Belarus, Kazakhstan, the Kyrgyz Republic, and the Russian Federation
- Initiate and review, with expert advice, drafts of laws and other legal acts
- Initiate parliamentary discussions of specific issues.

Both houses have committees on agrarian policy.[1] In the State Duma the 20-member Committee on Agrarian Issues is led by Alexei Chernyshev of the Agrarian Party. In the Federation Council the Committee on Agrarian Policy, with 13 members, is led by Valery Savchenko, governor of Belgorod *oblast* (region). Other members of that committee include Vasily Starodubtsev, governor of Tula oblast and leader of the Agroindustrial Union, or Rosagropromsoyuz (and a participant in the failed military coup of 1991), and Alexander Nazarchuk, head of the Altai Regional Legislative Body and a former minister of agriculture and food of the Russian Federation.

All issues concerning agrarian policy are discussed by both committees, separately and in some cases jointly. The views of the committees do not necessarily coincide. The Committee on Agrarian Policy is considered to be more liberal and (very slightly) more market oriented than the Duma committee; its members, who combine legislative work in the Federation Council with practical work as regional governors and heads of regional parliaments, tend to be more realistic. On several occasions the committee has stopped legal acts inspired by the State Duma, such as the most conservative version of the Land Code and the decision to increase the share of spending for agriculture in the federal budget to 15 percent.

---

1 The text and tables describe the situation at the time of writing, in September 1999.

The Agrarian Faction (the Agrarian Deputy Group) within the State Duma consists of 35 members of parliament (MPs), mainly from the Agrarian Party and the Communist Party of the Russian Federation (CPRF). The members of this group participate in the work of the Committee on Agrarian Issues. The Agrarian Faction votes jointly and solidly (100 percent) with the Communists. It also demonstrates a high level of voting solidarity with the Liberal Democratic Party of Russia (86.7 percent) and with Narodovlastie, or People Power (98.8 percent). The leader of the Agrarian Faction, former *kolkhoz* (collective farm) chairman Nikolai Kharitonov, is one of the most active antireformers among the MPs. His faction mounts strong opposition to any market-oriented decisions in the agrarian sphere.

Table 1. Political Parties in the State Duma, 1996

Political party	Leader	Number of seats	Share of total (percent)
Communist Party of the Russian Federation	Gennady Zyuganov	157	35
Independents (no party affiliation)		78	17
Our Home Is Russia	Viktor Chernomyrdin	55	12.5
Liberal Democratic Party of Russia	Vladimir Zhirinovsky	51	12
Yabloko Bloc	Grigory Yavlinsky	45	10
Agrarian Party of Russia	Mikhail Lapshin	20	4.2
Russia's Democratic Choice	Yegor Gaidar	9	2
People Power	Sergei Baburin	9	2
Congress of Russian Communities	Yuri Skokov	5	1.1
Forward, Russia!	Boris Fedorov	3	0.6
Women of Russia	Ekaterina Lyakhova	3	0.6
Other parties		15	3
Total		450	100

The Executive

The government of the Russian Federation is the main executive body. Its members are nominated by the prime minister, appointed by the president, and approved by the State Duma. There may be from two to eight deputy prime ministers (currently there are four), one of whom is responsible for agrarian issues. Under the direct supervision of this deputy are the Ministry of Agriculture and Food, the State Committee on Land Policy, the State Committee on Fisheries, the Federal Forestry

Service, and the Federal Hydrometeorology and Environmental Monitoring Service.

In addition to these bodies, various departments prepare issues for Cabinet meetings. Thus, the Department of Environment and Agroindustrial Sector Reform supports the work of the corresponding deputy prime minister.

The Ministry of Agriculture and Food is the main governmental body responsible for agrarian policy in Russia. The minister in charge usually has one or two first deputies and several ordinary deputies who handle the various areas of ministerial activity. The main departments of the ministry deal with economics and finance, land tenure issues, farm structure, state property, markets and food consumption, trade and foreign relations, and education and research. The ministry also includes such activities as state inspections, veterinary services, crop protection, and equipment maintenance. The State Committee on Fisheries and the Federal Forestry Service were formerly part of the Ministry of Agriculture and Food.

The Ministry of Agriculture and Food is the primary implementer of the federal agricultural budget. It elaborates the main annual programs in the agricultural and food sector and presents them to the Cabinet for approval. It coordinates its activities with the activities of the ministries of Economics and Finance, which also have agricultural and food departments. Together with the Ministry of Trade, the Ministry of Agriculture and Food participates in the World Trade Organization (WTO) process.

Under the umbrella of the ministry is an Advisory Agri-Food Council consisting of the leading agricultural people in Russia, from both academia and business. The council considers the major drafts of regulations issued by the ministry and makes comments, amendments, and recommendations. Its objective is to take into consideration broad public opinion on agricultural and food policy.

The agricultural bodies of the constituent members of the federation are administered both by the federal Ministry of Agriculture and Food and by regional governments, through regional departments or ministries of agriculture. These bodies are financed by regional budgets and have not been particularly responsive to the federal ministry.

The State Committee for Land Policy is responsible for land legislation, land engineering, and monitoring. It is currently in charge of land registration (several other bodies register nonland assets), but after

2000, in accordance with federal law, real estate registration will be shifted to the juridical system. The committee has branches in every constituent member of the federation and in every *rayon* (district). The federal budget finances this network. The committee is the only Russian governmental body funded by a special tax, the land tax.

Of the several bodies under the umbrella of the Ministry of Agriculture and Food that participate in formulating agricultural and food policy, one of the most questionable is the Federal Agency of Food Market Regulation (FAFMR). Created at the end of 1997 to replace the Federal Food Corporation, the FAFMR is a legal form of state unitary enterprise with the following tasks:

- Monitoring and evaluating agricultural production and food market conditions, providing information and support, and making market forecasts
- Promoting competition in the purchase and trade of agricultural products and food
- Preparing proposals on food market regulation for the Ministry of Agriculture and Food
- Carrying out trading and purchasing interventions for market stabilization, participating in the formation and functioning of wholesale markets, providing guarantees for transactions in food and agricultural production, and acting as state procurement agency for food deliveries to the governmental operational reserve and to the ecologically damaged territories.

One could predict that FAFMR would prove to be an easy and reliable channel for presenting ideas directly to and on behalf of the Ministry of Agriculture and Food. It is not surprising that its first and current director is Vladimir Loginov, the president of the Sugar Union of Russia, a producers' association.

Another such body under the Ministry of Agriculture and Food is the Special Fund for Preferential Loans for the Agroindustrial Complex, set up in 1997 to provide soft loans to agricultural producers (at one quarter the Central Bank interest rate). The fund was financed with budget resources and the previous debts of farms to the state. In 1997 its loans were disseminated by two commercial banks, SBS-Agro and Alfa-bank. In 1998 there was an open competition, and 12 banks were selected, including those two. Since the crisis of 1998, only SBS-Agro and Alfa-bank have actually been handling fund resources. The monies distributed through the fund are comparable in size to the federal agricultural budget, so decisions having to do with the fund are of great importance for agricultural and food policy in Russia.

Political Parties and Blocs

In the 1995 elections, 30 percent of the political parties and groups that took part had no position on agrarian issues (Table 2). Ninety percent of the electorate voted for parties that did have a policy.

Table 2. Agrarian Policy of Political Parties, Blocs, and Movements Participating in the State Duma Elections in 1995

Agrarian policy	Number of parties	Percentage of total	Percentage of electorate voting for party
None	13	30.2	5.75
Agrarian plank in party program	30	64.8	90.0
Nonsubstantive mention	12	27.9	12.4
Crucial importance	18	41.9	77.6
Party program includes party's own agrarian program or special agrarian issue	13	30.2	62.7

Note: *The term "party" includes political parties, movements, and blocs.*

The parties on the left, such as the CPRF, People Power, and the Agrarian Party, insist on an increased state role in regulation of the agrarian sector, on the assumption that the crisis in Russia's agroindustrial sector is the key reason for the macroeconomic crisis. The parties on the right, such as Russia's Democratic Choice, the Yabloko Bloc, Our Home Is Russia, and Forward, Russia! want to liberalize agrarian economics and trade and thus strengthen the position of the sector in the new market conditions. The Congress of Russian Communities takes a centrist position (Table 3).

The basic issues dividing the parties in the State Duma are landownership and the agrarian budget. The CPRF takes the orthodox Communist attitude that there is no right to private ownership of land, even to household and *dacha* plots (only "personal" ownership is excepted). This stance of the CPRF is the main reason that the new version of the Land Code, which simply repeats the articles of the

Constitution dealing with private ownership of land, is still pending. Even the CPRF's closest allies in the Agrarian Party do not support its position regarding landownership.

The Agrarian Party of Russia (AP) is the most influential party among the rural population. Its objectives are as follows:

- Rebirth of Russian villages
- Protection of the political, social, and economic rights and the legal interests of Russian farmers
- Protection of the food market and domestic agricultural producers
- Development of the agrarian sector on the basis of a multifaceted economic structure
- Promotion of true sovereignty of the people.

The constitutive forum of the AP was held in February 1993. The party's initiators and cofounders were the Agrarian Union of Russia, the Union of Agroindustrial Workers of the Russian Federation, the Russian Council of Kolkhozy, and former MPs of the Supreme Soviet of the Russian Soviet Federative Socialist Republic. The leader of the AP is Mikhail Lapshin, a former *sovkhoz* (state farm) director and a successful agrarian lobbyist. The party's backbone consists of former Soviet and Communist *nomenklatura* (bureaucrats) in the agrarian sector, kolkhoz and sovkhoz directors and their deputies, and local administrators. The strata of representatives of ordinary employees of agricultural enterprises, private family farmers, owners of household plots, and other sectors of the rural population are very thin in the AP. The party has a three-tier structure, with federal, regional, and local/primary levels; there are 80 regional branches (out of a possible 89). The AP has a regular newspaper, *Russian Land*, which has been published since April 1996, and several regional newspapers. It maintains contacts with foreign allies such as the Rural Party of Ukraine, the Agrarian Democratic Party of Moldova, the Farmer Party of Poland, the Bulgarian Peasant Union, and the German Peasant Union. In August 1996 it joined the electoral bloc called the People's Patriotic Union of Russia.

Table 3. Agrarian Programs of Political Parties, Movements and Blocs Represented in the State Duma

Party or movement	Agrarian plank in the party program	Statement of the priority of agricultural development	Support for strengthened role of the state in the regulation of agriculture	Positive attitude toward private ownership of land	Positive attitude toward purchase and sale of plots of land
Communist Party of the Russian Federation	+	+	+	—	—
Our Home Is Russia	+	+	+	+	+
Liberal Democratic Party of Russia	+	+	+	+	***
Yabloko Bloc	+[a]	—	—	+[b]	+
Agrarian Party	+	+	+	+	***
Russia's Democratic Choice	—	—	—	+	+
People Power	+	—	+	*	**
Congress of Russian Communities	+	+	+	+	***
Forward, Russia!	+	—	—	+	+
Women of Russia	—	—	—	+	***

a. The bloc's agrarian program was presented in May 1999; previously, Yabloko had not had an agrarian program or a special agrarian plank in its political program.

b. Yabloko insists that the "land question" has to be decided on the regional level.

* Private ownership of land to be restricted to household, dacha, and garden and orchard plots.

** Purchase and sale of land to be restricted to household, dacha, and garden and orchard plots.

*** Heavy restrictions on purchase and sale of land.

Social Organizations

Noncommercial social organizations are not permitted to have political aims in their charters. The most influential such organizations are discussed in detail below. A number of other agrarian groups were organized to fit the current political situation and do not have much impact on social and political life in Russia. They include the Agroindustrial Union, the Russian Union of Rural Women, the Russian Agrarian Union of Youth, the Russian Peasant Foundation, the Union of Landowners, the Movement of Women Farmers, the Union of Small Agricultural Producers, and the Union of Rural Credit Cooperatives.

Rosagropromsoyuz

The Agroindustrial Union of Russia (Rosagropromsoyuz) was established in April 1997. Its elected chair is Vasily Starodubtsev, governor of Tula oblast. The organization's members include representatives of agricultural enterprises, processors, storage and transport enterprises, machinery stations, producers of fertilizers and agricultural chemicals, and service and scientific organizations. The majority of its membership is made up of representatives of large agricultural enterprises. (Only two private farmers are members.) Rosagropromsoyuz's predecessor, the Agrarian Union, consisted of legal entities in the agricultural and food sector. It was created in 1990 to lobby for the interests of "red" agricultural directorates in the government and to prepare for the first parliamentary elections. Its role decreased after the establishment in 1993 of the Agrarian Party, to which the union gradually delegated most of its political functions. In 1995 the Civil Code was enacted, and since the Agrarian Union's charter was in conflict with the code's provisions on social organizations, which were to consist only of physical persons, the organization was superseded by Rosagropromsoyuz.

The main chartered goals of Rosagropromsoyuz are to:

- Promote the creation of the economic, social, and legal conditions for the successful development of the agricultural sector

- Develop financial and credit infrastructure for the agroindustrial complex

- Support price parity between agriculture and other branches of the economy, improve the tax system in favor of agriculture, and develop scientific support for the agroindustrial complex.

Every year, Rosagropromsoyuz signs an agreement with the government of the Russian Federation. This agreement, however, is merely declaratory, and the provision stating that Rosagropromsoyuz is to receive budgetary funds for its activities has never been carried out.

Rosagropromsoyuz has branches in all constituent members of the federation except Chechnya. The main body of Rosagropromsoyuz, the forum, meets at least once every three years.

AKKOR

The Association of Private Farms and Cooperatives of Russia (AKKOR) was created in 1990 on the initiative of the new stratum of private farmers. Unlike Rosagropromsoyuz, AKKOR has a progressive and democratic image. Since the *perestroika* era, it has been considered one of the most market-oriented organizations in agriculture and has been lobbying for private ownership of land. Its leader is Vladimir Bashmachnikov, an MP in the State Duma. AKKOR has 850 first-tier local organizations in 68 regions of Russia. Its main goals include protecting the civil, political, economic, and cultural rights and freedoms of private farmers; promoting agrarian reform and private initiatives for development of the agrarian sector; and supporting the rule of law in the design of the legislative base relating to land and agriculture. To achieve these goals AKKOR takes part in launching proposals for the government, participates in discussions on the agrarian budget, and maintains contact with political parties and movements and with religious and social organizations regarding the social and economic development of Russia, continuation of economic reforms, and the sovereignty of Russia. The association's main body is the forum, which meets not less than once a year. AKKOR is a member of several international organizations, including the International Federation of Agricultural Producers.

At the beginning of its existence, AKKOR was involved in some economic activities: the first financial support for private farmers was disbursed via its branches, and an attempt was made to organize cooperative banks and an insurance company. These efforts failed, and now AKKOR minimizes its economic activity. AKKOR sponsors "Russian Farmer," an agricultural fair for family farms that has been held in St. Petersburg during the last week of August every year since 1992. This fair is regarded as an important event; many top government officials, as well as representatives of the governments of other countries and international organizations, participate in it, and many decisions and announcements are made in the context of the fair.

Business and Producers' Associations

Business and producers' associations have begun to emerge following the privatization of the food-processing sector. The most important of these are discussed below. Formation of other business associations, including a tea union and an alcoholic beverages producers' association, is under way.

The Grain Union of Russia

The Grain Union of Russia, established in 1994, was the first of these associations. It was organized by the federal corporation Roskhleboproduct, the joint-stock company Exportkhleb, the corporation Mosoblkhleboproduct, OGO (one of the first private grain-trading companies in Russia), the Russian Grain Exchange, the Moscow Trade Exchange, and other entities. Alexander Yukish is its president, and Arkady Zlochevsky, the president of OGO, heads the board of directors.

The main chartered goals of the Grain Union include interacting with the government on issues concerning grain market regulation; elaborating and reviewing laws and instructions related to the grain business; and providing members of the union with information on trends in the grain market. The Grain Union is open to new members who are legal entities involved in grain purchase, storage, processing and trade, commodity exchanges, and other aspects of the grain trade.

The Sugar Union

The Sugar Union was organized in 1996 for "the effective solution of the problems of the sugar industry and sugar market of the Russian Federation; representation and protection of sugar plants and operators of sugar market interests in governmental structures, social organizations, and other organizations including foreign ones; strengthening of united actions and mutual assistance." The main cofounders were the corporation Russian Sugar and the joint-stock companies Lipetsk Sugar, Tula Sugar, Rossakhar, and Alfa-Eco. The Sugar Union represents 96 sugar plants throughout the country. Vladimir Loginov, former director of Russian Sugar and now director of the FAFMR, has been the group's president since 1997.

The association's goal is to establish a sugar regime that favors domestic sugar producers. The introduction of high import tariffs in 1997 was the direct result of its lobbying. The Sugar Union has agreements with many sugar unions elsewhere, the most important one being that with its counterpart in Ukraine, Russia's primary competitor in low-cost sugar production.

The Meat Union

The Meat Union was organized in November 1998 on the initiative of the Committee of Economic Security of the Meat and Meat Products Market and under the auspices of the Trade and Industry Chamber of the Russian Federation, the Association of Meat Processors of Russia, the Veterinary Association of Russia, the trade and industrial companies Rosmyasomoltorg and Rosmyasomolprom, the agroindustrial enterprise Cherkızovsky in Moscow, the share company Samson in St. Petersburg, and other well-known organizations. The Ministry of Agriculture and Food of the Russian Federation supported the establishment of the organization. The members are 272 enterprises representing 72 regions. Iosif Rogov, rector of the Moscow State University of Applied Biotechnology and academician of the Russian Academy of Agricultural Sciences, is the elected chair.

The main task of the Meat Union is to promote favorable conditions for domestic producers and processors, to develop the infrastructure of the meat market, and to eliminate unscrupulous competition between domestic and foreign companies. The Meat Union intends to present proposals to the Russian government regularly. Among its first proposals were to reduce import tariffs on raw meat for the processing industry, since imports have taken on critical importance following the August 1998 financial crisis.

Research and Consulting Organizations

Research and consulting organizations influence the development of agrarian policy to some degree. Research institutes in the agricultural sphere work under the auspices of the Russian Academy of Agricultural Sciences (RAAS). The RAAS unites 199 research institutes, 24 experimental agricultural stations, 47 genetics and biotechnological centers, and a central agricultural library. The presidium of the RAAS is a nest of the ultraradical left opposition; most of its members belong to the CPRF.

The Institute of Agrarian Issues and Informational Sciences (the Agrarian Institute), founded in 1991 by the president of the Soviet Union Academy of Agricultural Sciences, Alexander Nikonov, is the most liberal of the RAAS research institutes. It was responsible for drafting the basic concepts and legal acts initiating agrarian reform in Russia. Two other economics institutes of the RAAS—BNIESKh and BNIETUSKh—represent the views of the agrarian opposition and draft RAAS memos to the government on agricultural and food policy issues.

The RAAS has a network of economics institutes in the regions, and some of them have considerable influence on agrarian policies at that level. For example, the Novosibirsk, Rostov, and St. Petersburg institutes participate in the formulation of regional agrarian policies.

The network of academic institutions for economics includes several centers for agricultural research, some of which significantly influence federal or regional policy. The agricultural division of Yegor Gaidar's Institute for the Economy in Transition (IET) produces many recommendations for policymakers. IET has recently established an Analytical Center for Agri-Food Economics. The Saratov Institute for Agricultural Economics is a regionally important research body.

The Foundation for Support of Agrarian Reform and Rural Development (Rosagrofond), established in 1997, is well known for the development and dissemination of the Nizhny Novgorod land privatization and farm reorganization model. Rosagrofond provides legal and economic consultations for federal, regional, and local administrations, farm directors, and other participants in agrarian reform. Similar consulting foundations were created in the Orel, Nizhny Novgorod, Rostov, and Volgograd regions. Foundations for promotion of agrarian reform are united in the Association of Agrarian Funds (AGRO), established in 1999.

Regional Agricultural and Food Policy

Since the beginning of the reforms, Russia's agricultural and food policy has tended to be more and more regionally based. Thus, at the start of the reforms the federal budget financed around one third of total spending for agriculture; regional budgets funded the remaining two thirds. In 1998 the share of the federal budget for agriculture was only 20 percent. But even when the money comes from the federal level, regional governors tend to concentrate its distribution in their own hands.

Since 1993, regional governments have maintained the livestock subsidies, the second largest program in agriculture. The agricultural and food reserves that regional authorities amass for "regional purposes" exceed by several times the federal reserves. These reserves are an important tool for influencing agricultural and food industries, and many net exporting regions impose trade bans and barriers to trade in agricultural and food commodities in order to maintain their reserves. Such activity is counter to Russia's Constitution, but the federal power has no leverage for stopping these interregional trade restrictions.

The significant rights that regions exercise in the schemes of privatization of agriculture and the food industries have a great effect on agricultural and food policy in the country as a whole. Some regional governments maintain price regulations for both the retail and the agricultural markets.

Regionalization of agricultural and food policy increased after the crisis of 1998. Faced with a sudden demand for foodstuffs, many regional authorities introduced price regulations, trade barriers, and controls on the retail network.

The regional governors are among the most important driving forces in Russia's agricultural and food policy today. Constituent members of the federation have formed regional associations and unions such as Siberian Concord, the Ural Union, and the Association Center. Most of these associations have their own agrarian programs. Thus, Alexander Nazarchuk, the former federal minister of agriculture and food, now governor of Altai *kray* (territory), is the head of Siberian Concord's Agrarian Committee. The upper house of Parliament is another forum within which governors can discuss and influence agrarian policy, especially given the growth of the power of that house.

Conclusions

In the course of reform in Russia, a new system for formulating agricultural and food policy has emerged. It is still very complicated and nontransparent and does not fully respond to the demands of an open, democratic society. However, the main driving forces are emerging: the legislature, governmental bodies, and political and public organizations. The development of these bodies is uneven; for instance, representation of the interests of businesses in the agricultural and food sector has just begun.

Russia's federal structure makes for a problem of imbalance between the federal and regional levels of decisionmaking. There is still a need for cooperation among regional authorities and the federal government on the regulation of the agricultural and food sector.

The contradictions in the process of formulating agricultural policy stem from the lack of public consensus regarding the fate of the agrarian reform, as is shown by the diversity of the agrarian programs of the various political parties. Even the extreme positions are supported by similarly aligned research and consulting bodies, which enlarges the gap between left and right positions.

Nevertheless, as the economic process promotes a greater consensus on the situation in the agricultural and food sector, the practical interests of economic agents and the emergence of pressure groups will drive policymakers toward a more sustainable policy. The public interest will inevitably demand that the policy-forming process become more transparent and straightforward.

References

OECD (Organisation for Economic Co-operation and Development). 1998. "Monitoring of Agrarian Policy: Russian Federation." Paris.

Russian Federation. 1997. "On the Organization of the State Unitary Enterprise 'Federal Agency of Food Market Regulation' under the Auspices of the Ministry of Agriculture and Food of the Russian Federation." Government Regulation N 1224, 26.09.97. Moscow.

Land Reform: Expectations and Social Consequences

Vladimir Bogdanovsky

Reform of the social order in Russia's rural areas has had as its premise an idea almost primitive in its simplicity: that the distribution of the "people's property" to natural and legal persons would, virtually overnight and almost by itself, bring about fundamental changes in the agricultural sector. It would eliminate alienation of the worker from labor, create new incentives for the economic behavior of the worker, and eventually increase productivity. Production of agricultural commodities would be invigorated and its stable growth ensured. The result would be a higher standard of living for agricultural workers, social and economic restoration of rural areas, an adequate supply of domestic farm produce at competitive prices, satisfaction of consumer demand, and an increase in domestic demand for Russian agricultural products.

The attempts to implement the original idea of institutional changes in forms of landownership showed a remarkable consistency in the reform years. Land reform has been given credit for dismantling the state's monopoly of land as property and as an economic object. The state now owns only about 17 percent of agricultural land; most of the remaining land is privately owned by collective farms or by individuals. The structure of ownership of other means of production in agriculture has changed in a similar manner.

By now, it would seem, one might expect to see evidence of the achievement of the strategic objectives of reforming the rural social setup: a real increase in the well-being of the rural population and saturation of the national food market with domestic produce. Neither has happened. As sober-minded scientists and politicians had warned, agricultural production has dropped, and we have witnessed a drastic decline in the standard of living in rural areas, disintegration in the social sphere, and a shrinkage of the rural population as a result of migration from the countryside.

Even the authorities now recognize the destructive effect of forcing agricultural production into the private mode, predominantly in the form of individual farms. The land reform policy failed to allow for the socioeconomic conditions that had evolved in Russia's agriculture, and it was never supported with appropriate macroeconomic measures or the creation of market structures. Policymakers overestimated the market's ability to self-regulate in the context of reduced government involvement in agricultural production.

Farmers' Preferences

Among the widely discussed errors, mistaken assumptions, and shortcomings of the rural reform, a set of key factors has so far drawn little attention: rural potential for social reform, farmers' value preferences, and the extent to which farmers are prepared to act as agents in the institutional change of ownership, especially in ownership of land. These seemingly minor considerations may frustrate reform efforts or at least considerably diminish their impact, given the well-known conservatism of the farming community and its lack of responsiveness to sudden "movements." A recent study of farms with various forms of ownership in four *oblasts* (regions)—Vladimir, Samara, Tambov, and Moscow—sheds some light on this matter.

Leaving private farmers aside, the land reform has mostly been reduced to official "distribution of land shares." There is abundant evidence that the process fails to transform new landowners into fullfledged agents of land relations because it leaves the economic mechanism of land relations virtually intact. Could that be why only 1 of 10 employees at the surveyed agricultural entities supported the reforms, while three times as many respondents expressed strongly negative views about the changes? (The remaining 60 percent was undecided.)

Agrarian policies are increasingly focusing on private ownership of land, as evidenced by a recent decree, "On Implementation of the Constitutional Rights of Citizens to Land." The decree once again authorizes redistribution of land in favor of private ownership; indeed, it allows land shares to be carved from collectively owned land without the consent of co-owners. It will, however, take considerable interference "from above" to significantly invigorate land turnover in agricultural enterprises. Proper prerequisites for changes in ownership of land within agricultural communities have yet to evolve. In our 1995 survey of agricultural entities under various forms of ownership, four out of five employees said they intended to keep their land in collective use (Table 1). Only 11 percent of the respondents had other intentions, including 4 percent who planned to lease their land and another 4 percent who intended to increase the size of their subsidiary farms.

These average values hide considerable differences among individual agricultural entities. The differences are determined not so much by specific forms of ownership as by the preform level of socioeconomic development of the agricultural entity. There may be a tendency for a growing number of employees to want to transfer their land shares for collective use by economically sound enterprises.

Table 1. Intentions of Employees of Agricultural Entities Concerning Use of Their Land Shares, 1995

(percentage of the total number of respondents)

Intention	Average for surveyed agricultural entities	Variations across agricultural entities
Leave in collective use	78	59–94
Sell	1	0–3
Lease	4	0–13
Increase the size of private residence	4	0–9
Start a private farm	2	0–3
Uncertain	11	2–27

It would be unrealistic to expect a different set of preferences that would favor more active land turnover in the countryside. Farmers remain cautious about the very basis of land turnover—institutional changes in the ownership of land. In other words, the rural population has not been reform minded enough for private ownership of land to advance radically. Attempts at artificially prodding the process forward have brought about results the opposite from those intended, as evidenced by our surveys and polls in recent years.

A 1990 poll indicated that 16 percent of employees of agricultural entities supported introduction of private ownership of land as the legal basis for economic activities; another 36 percent said they would accept the innovation but only for their subsidiary farms, private housing projects, and similar purposes; 40 percent strongly opposed private ownership of land; and 8 percent was not sure. Thus, in the initial stages of agrarian reform, over half the employees of agricultural enterprises endorsed private ownership of land to some extent. By 1995, however, the percentage of those polled who supported private ownership had dropped to 32 percent. Apparently, the spontaneous and ill-thought-through nature of the reforms, coupled with their lack of social focus, had diminished the reform potential in rural communities in many regions by 50 percent.

At the same time, significant changes have occurred in farmers' views on private ownership of land, with the share of respondents who unconditionally welcome it plummeting to 7 percent. Even proponents of private ownership of land (20 percent of the respondents) believe that private ownership should be delayed because of the lack of an adequate legal environment and the high probability that bureaucrats will "privatize" land for themselves for resale purposes. It is notable that only 60 to 70 percent of the private farmers surveyed support private ownership of land as the sole legal basis for economic activity; the others see leasing as an acceptable alternative. Only 1 out of 20 respondents found it important to own the land occupied by their subsidiary farms—and this when the issue has already been resolved.

One reason for the low acceptance of private ownership of land used for subsidiary farms is that, except in the vicinity of large cities, turnover of such land is insignificant. In addition, the land tax rate is low. As a result, most rural people do not realize that they are managing their subsidiary farms in the context of a radically new form of landownership.

Expectations and Current Perceptions Regarding Land Reform

The low level of reform expectations is manifested in the rural population's estimate of the potential opportunities inherent in the reform of land relations and economic activity (Table 2).

Table 2. Agricultural Employees' Assessment of the Effect of Land Reform on the Productive Environment and on Socioeconomic Processes

(percentage of all respondents)

Activity	Expect the situation to improve	Expect the situation to deteriorate
Use of land	16	23
Use of technology	10	24
Use of manpower	12	27
Labor incentives	11	14
Output of farm produce	16	21
Buildup of social infrastructure and facilities	10	21
Ecology of land use and environmental protection	9	26
Development of plots of land adjacent to residences	40	8
Living standards of rural population	6	25
Social security net	7	29
Morals and cultural level of the rural population	8	27

The share of respondents expecting improvements in the production and labor areas rarely exceeds 10 to 15 percent. The remainder does not simply consist of those who have no positive expectations with regard to the reform and believe that the status quo will remain unchanged. Were that the case, the situation could be evaluated as being positive at this early stage of agrarian reform. But in fact, one of four "neutral" farmers sees a negative and destructive potential in the reforms, especially regarding use of manpower. (Their assessment of the impact of reform on labor incentives and land use is less negative.)

Expectations regarding the social sphere and socioeconomic implications are even lower. Only 8 to 10 percent of the rural population harbors hopes for a positive outcome. And while the ratio of negative to positive expectations was 1.5 to 2 in the production sphere and 2 to 1 as regards labor, the corresponding ratios for the social sphere and for socioeconomic implications are almost twice as large.

Even the extremely low level of expectations among farmers exceeds the actual changes in land relations and forms of ownership and is therefore higher than the estimate of such results by employees at agricultural entities. For instance, to evaluate the use of agricultural land, one can look at the reduction in productivity of most crops, which is 30 to 50 percent. Only 1 out of 10 respondents noted an increased interest in more effective use of land, and a third of the respondents pointed to a decrease in the motivation to use land effectively. Evaluations regarding the ecology of land and environmental protection are even lower. A negligible 2 percent saw positive changes in these areas, while two thirds of the respondents mentioned negative dynamics. An overwhelming 82 percent cited depletion of resource inputs (hardware, fuel and lubricants, fodder, and so on), and virtually no one said that inputs had increased.

Labor Incentives

Even farmers' moderate expectations for social and labor relations have not been met in full. On the whole, the significance of labor incentives has decreased, as corroborated by the survey results; 8 percent of respondents mentioned increased wages, but 67 percent said wages had declined. Other factors of labor activity have dropped by a similar proportion. Only 1 in 10 respondents has greater labor incentives; 1 in 3 has seen labor incentives drop.

The significance of the land reform and of the changes in the rural economic structure is vitiated by the lack of tangible progress toward overcoming alienation from labor. Less than 2 percent of agricultural employees felt that reform had given them greater opportunities to act as economic agents, and 21 percent said that they were less in a position to do so as a result of the reform. Thus, among the 11 percent of respondents who had expressed optimistic expectations, only one out of six had seen these expectations fulfilled, while pessimistic expectations (expressed by 14 percent) were exceeded, in the event, by 50 percent—things turned out worse than expected.

Living Standards and Morale

The situation is no better regarding living standards or the social infrastructure in rural communities. How could it be better when agriculture is at the bottom of the wage scale, paying employees 2.5 less than industrial operators? (The difference was only 7 percent as recently as five years ago.) The gulf in social payments between industry and agriculture has also widened; social payments in industry exceed those in agriculture by 300 percent.

Living standards are also being eroded by fast-growing unemployment. As of late 1994, official statistics put unemployment in rural areas at 2.5 percent of the able-bodied population, but that figure does not include the approximately 10 percent of agricultural employees who suffer from hidden unemployment.

Fewer investments are being made in the rural social sphere: investments in this area by the agroindustrial complex dropped by almost 50 percent in 1991–95 in comparison with the previous half-decade. The amount of new housing dropped proportionately, the number of newly built general-education schools decreased by 2.2 times, and the number of preschool facilities fell by 3.5 times. During the same period 46 constituent regions of the Russian Federation failed to put into operation any new hospitals, and 34 commissioned no new recreational facilities. Over the past few years about 7,000 kindergartens and 500 schools have been shut down, along with 3,000 recreational buildings and 1,200 medical establishments. There has been a 50 to 90 percent decrease in the number of catering, service, and trade outlets.

What we are seeing is the disintegration of the rural social sphere. It is not surprising that 80 percent of the respondents said living standards had decreased and that 60 percent cited reduced availability of services and poorer performance by medical establishments, schools, libraries, and sports facilities. It appears that even the insignificant positive impact that farmers expected the reform to have on their living standards and on the availability of services (see Table 2) was achieved only partially (30 to 50 percent).

Feelings of disillusionment and despair have grown stronger. About two-thirds of the respondents mentioned declining moral standards in their communities, and four-fifths said that drinking and crime were on the rise. Eighty-one percent said they had less confidence in the local authorities, and 85 percent said the same with regard to the federal government.

Progress on Subsidiary Farms

Of all the spheres and processes that might be positively influenced by the changes in land relations and in the economic system, the subsidiary farm experienced the largest positive impact. As Table 2 shows, 40 percent of the employees of agricultural entities saw the reform as potentially positively for the development of their subsidiary farms. (The remainder saw no such potential, and 8 percent believed that the reforms would have a negative impact.) Reality outperformed expectations by 50 percent; 60 percent of the respondents noted a higher employment rate and greater incentives for managing their subsidiary farms. Moreover, 5.6 percent of the 7 percent of farmers who focused on private enterprise and self-employment saw their subsidiary farms as primary employment (the share varies between 1 and 14 percent across agricultural enterprises), and less than 1 percent said they planned to make full-fledged private farms their primary form of employment.

Relationship to the Parent Enterprise

The growth of the subsidiary farm in recent years and its emergence as an economic agent furnish a visible expression of the shift of labor resources from agricultural employers to subsidiary farms. The farmers use the resources of their sponsors in their own production, bleeding the sponsors white. The paradox is that the sponsoring enterprises, strangled in many respects by current agrarian policies, have been encouraging and assisting their employees to build up their subsidiary farms.

This phenomenon stems from the deteriorating economic situation of the majority of agricultural operators as a result of growing price disparity, an excessive tax burden, nonpayments, and other effects of the sociopolitical and economic crisis. In consequence, wages in agriculture are the lowest in the economy, and people go unpaid for months at a time. The work collectives at agricultural operators (which in most cases are interested in helping their employers stay afloat) have been seeking a way out through payment in kind and through various types of assistance to employees' subsidiary farms. This assistance has begun to embrace the area of land relations. In addition to being given access to fieldlots, hayfields, and pastures, subsidiary farms have recently been allowed to increase in size at the expense of employees' shares in the employer's land.

Importance of Subsidiary Farms

Increased inputs have resulted in higher production by subsidiary farms; their share in national agricultural output climbed from 24 percent in 1990 to 38 percent in 1994, while the share of collective farms' output dropped from 76 to 60 percent. The gain in the significance of subsidiary farms has been achieved primarily through a dramatic increase in the number of livestock. Our studies in Elkhov district, Samara oblast, indicate that, on average, the number of livestock owned by subsidiary farms is about 50 percent that of the collectively owned livestock on specialized livestock-breeding farms. A new dam-based type of subsidiary farm has begun to evolve, and the scale of its livestock operations, unattainable by most private farms, is impressive. The average subsidiary farm owns almost 2 cows, 1.2 breeding sows, and 2.3 ewes. The total number of livestock is two or three times the number of dams. All subsidiary farms breed poultry and horses (a particularly important point, given the currently high prices of hardware, fuels, and lubricants).

Subsidiary farms within the surveyed agricultural operators meet only 9 percent of their demand for fodder through plots adjacent to their residences. The balance is provided by the land of the collectively owned operator. Thus, considerable redistribution of land in favor of subsidiary farms is taking place through numerous channels, although it is not officially recorded. There is also a tendency to maintain the status quo. Indeed, 21 percent of rural dwellers would like to see an increase in the size of their subsidiary farms, but only 5 percent of them intend to enlarge their farms officially and legally, at the expense of their land shares.

Attitudes toward Labor and Forms of Ownership

The emergence of a new set of work incentives shaping the economic behavior of rural residents constitutes the driving force behind the achievement of most of the objectives of the reform of production relations in the countryside, including the strategic social goal of improving the well-being of the people. The development of such incentives depends on institutional changes in property relations, yet only a fraction of agricultural workers has an adequate understanding of these changes. Logically, it is among this fraction (in a real-world situation corresponding to their value preferences and to changes in property relations and economic activity) that a new mechanism of incentives and increased labor activity is evolving.

For instance, a relatively small number of agricultural enterprises, which boasted a high level of socioeconomic development before the crisis, managed to survive and have even—by deepening intraenterprise organizational and economic relations—incorporated certain elements of more advanced and efficient property relations and of the market. Among the heads and employees of such enterprises, 70 percent of the respondents make full or above-average use of their labor potential. In the bulk of agricultural enterprises however, where the reform became stalled at the stage of formal "distribution of land shares," that indicator is 15 percentage points lower.

There is also considerable differentiation in the structure of labor motivation. For example, among farmers and employees of agricultural enterprises in the first (that is, successful) group of enterprises, the number of people focusing on wages and on the content of work exceeds the corresponding numbers in the second, "stalled" group 3.7 and 3 times, respectively.

Among farmers, 9 in 10 felt that they were owners of their businesses and derived greater job satisfaction, and 100 percent had become more active in their work. Agricultural operators in the first group, too, stand out in certain aspects. For instance 1 in 10 of their employees indicated an increase in labor discipline and an improved work attitude. Such evaluations should be viewed against the background of negative evaluations, which in the first group are 3 to 10 times lower than for the bulk of enterprises.

The "skidding" of the reform, which has resulted in a severe crisis, should be attributed not only to the lack of appropriate macroeconomic measures and the failure to build market structures but also to overestimation of the market's ability to self-regulate. Of vast importance are farmers' value preferences, their potential for reform, and their ability to accept institutional changes in property. Those factors, which have yet to be properly taken into account, are essential for launching the phase-out of collective and state-agricultural enterprises and accelerating the turnover of land shares.

The reality of the situation calls for better balanced, differentiated, and targeted action. In particular, in line with the actual socio reformist potential of rural areas in many regions, the land reform and agricultural management should take into account the following findings.

1. Only an insignificant proportion of employees of agricultural enterprises (2 to 5 percent) sees private ownership of land, implemented via self-sufficient subsidiary or private farms, as a basis for boosting economic behavior and labor activity.

2. A rather meaningful proportion of employees (8 to 21 percent) prefers both private and collective ownership implemented on the basis of small-scale production labor units that enjoy a high degree of economic independence and responsibility while operating within a collective enterprise.

3. The remaining agricultural workers (75 to 85 percent) are not concerned about specific forms of ownership; some of them would choose small-group economic activity with limited economic independence, while others support modified traditional large-group economic units.

It must be recognized that transformations in land relations and forms of economic activity in accordance with the value preferences of the second group of employees, and partly of the third group of employees, are largely being hampered by lack of knowledgeable and active employees who are capable of initiating and leading the reforms. Selection and training of potential leaders have to be addressed to move the reform process forward.

Public Opinion concerning Russia's Agrarian Reforms[*]

Eugenia Serova

Russia's agrarian reforms were launched in late 1991 with the adoption of a set of decisions on farm restructuring and land privatization, as well as privatization and decentralization of downstream sectors such as the food industry and the retail network.

The elements of the agrarian policy were controversial, and the policy was constructed largely through ad hoc decisions. There has been an obvious movement away from the centrally planned economy but no integral policy as yet, and the heritage of the previous economic paradigm still influences policymakers' views. Unlike the situation in most countries of Central and Eastern Europe, political consensus regarding the reforms is missing in Russian society, with detrimental consequences for the restructuring of the agricultural sector. After eight years of agrarian reform, the majority of farms are insolvent and not fully market oriented, markets are rather inefficient, rural areas are deteriorating, and policy is inconsistent.

Other chapters of this volume describe the reforms and discuss the reasons for the modest results. This chapter deals with the perceptions and opinions of influential strata in the agricultural sector and related fields. The information was collected from a survey conducted in 1999.

More than 500 questionnaires were circulated among experts in eight *oblasts* (regions) of Russia: Moscow, St. Petersburg, Pskov (northwest), Rostov (south), Orel (central non-black-soil area), Voronezh (central black-soil area), Tomsk (western Siberia), and Irkutsk (eastern Siberia). One part of the sample consisted of federal and regional legislators and officials, one part of agricultural and agribusiness managers, and one part of academics, extension service experts, agricultural journalists, scientists, and selected others. About 45 percent of the sample was between 35 and 50 years of age; a quarter was younger than 35. The breakdown by gender was about 50–50. The political spectrum within the sample was rather representative: 12 percent identi-

---

[*] The article is based on the results of a study initiated and financed by the World Bank (L.A. Norsworthy, Knowledge Manager) and carried out by a research team consisting of Eu. Serova, N. Karlova, I. Kramova, S. Kramova, O. Pysmennaya, and T, Tikhonova.

fied themselves as on the right, 26 percent as center right, 28 percent as center left, and 10 percent as left; 24 percent did not have a political affiliation (see Annex Table A-1).

The respondents were asked about their agreement with the reforms, their perceptions about implementation of the measures, and their evaluation of the components of the reforms. Another set of questions concerned the social and political aspects of the reforms, including their effect on rural standards of living. Finally, the experts were asked about the prospects for Russian agriculture.

Attitudes toward the Reforms

The experts surveyed were asked for their views on five components of the reforms:

- Land privatization
- Farm restructuring
- Privatization of the food industry
- Emergence of market infrastructure
- Current government agricultural and food policy.

Respondents were queried as to whether they thought the reform was necessary (Table 1), whether they thought the concept had been implemented (Table 2), and whether they agreed with the concept behind the reform Table 3). They were asked to evaluate, on a five-point scale, the results of each component of the reforms and of the reforms in general.

Land Privatization

An absolute majority (almost 77 percent) of the sample considered land reform crucial for the further development of agriculture in Russia. The share of officials who support land reform was even higher: 78 percent. Farm managers and academics are more conservative (71–72 percent). As might be expected, respondents from the right side of the political spectrum were more likely to approve of the land reform (80–83 percent); 73 percent of those who identified themselves as center left and only 60 percent of those on the left supported the reform. Those under age 35 were more supportive than the oldest (over age 50) segment, with approval rates of 85 and 72 percent, respectively. Among the sampled regions, Orel and Tomsk were least convinced of the need for land reforms (63 and 64 percent).

Table 1. Responses to the Questions Regarding the Necessity of Reforms

(percent)

Question	Yes	No
Was land reform necessary?	76.9	22.3
Was farm restructuring necessary?	59.7	39.9
Was privatization of the food industry necessary?	52.0	45.0

Among all the respondents, 29 percent said land reform had not been implemented and 48 percent said it probably had not been. Thus, almost three quarters of the respondents thought land reform had not been carried out. Only 8 percent said reform had been implemented and 15 percent said it probably had been. The most pessimistic group was the farm managers (84 percent negative); the most optimistic views were held by the mass media (71 percent negative). Roughly a third of the right and center right respondents were convinced that land reform had been carried out. Negative views were expressed by 82 percent of those on the left and by 73 percent of the center left. Half of the young experts but only a quarter of the oldest experts said the reform had been carried out.

Table 2. Responses to Survey Questions about Implementation

(percent)

Question	Yes	Probably has been	No	Probably not
Has land reform been carried out?	8	15	29	48
Has farm restructuring been carried out?	15	23	39	23

Among those who said land reform had been implemented, less than a third were not in agreement with the reform concept; those in agreement included most of the professionals, half of the mass media, almost all those on the right, and 60 percent of those on the left. No one among those who said reform had been implemented absolutely rejected the concept.

It is notable that in Pskov oblast only 56 percent of the experts answered that land transformation had not taken place there. That could be a reflection of the low efficiency of the reforms there. Pskov and Orel had the highest level of agreement with the concept of the land reform

(64 and 71 percent, respectively) In Moscow 33 percent of those who said that reform had occurred agreed with the concept.

Table 3. Responses on Agreement with the Reforms
(percent)

Question	Agree	Agree more than disagree	Disagree more than agree	Disagree
Do you agree with the concept of land reform?	1	6	3	90
Do you agree with the concept of farm restructuring?	12	30	45	13
Do you agree with the concept of food industry privatization?	5	22	35	36

The score for evaluating the land reform (on a scale from 1 to 5, with 5 the highest) averaged 2.0. There was almost no difference between the left (1.4–1.6 points) and the right (2.2). Among the regions, Pskov experts gave the highest estimate (2.3, on average), which is surprising, given the background of agricultural failure in the region during the reforms.

It is obvious from these results that the main part of the agricultural establishment looks forward to land reform but that agreement on the concept behind the reform is lacking. Since certain land transformations have in fact taken place during the past eight years, it seems that those who said reform had not been implemented (or had probably not been implemented) simply do not share the concept. Thus, those on the left, who are extremely unhappy with the land transformations and were in large part not enthusiastic about land reform, were most likely to say that the reforms had not been implemented. The same is true of the oldest experts. This could be the background for the dissatisfaction expressed by these groups (they do not want reforms, and reforms do not occur), but their perceptions do not match real-world observations.

The high level of conservative attitudes among academics and extension system experts helps keep reforms in abeyance; these groups often formulate concepts for policymakers.

There is a similar divergence of opinion on the introduction of land property rights and land transaction rights. Many experts saw these as either the most positive result or the most negative result of land transformations. Among the positive results mentioned were the growth in legal awareness of the peasantry, the increased consciousness of property rights, and the provision of sufficient food supplies in crisis circumstances. The negative evaluations break down into two groups. The first group is founded on a negative attitude toward the reform and includes such points as the decline in agricultural production and the deterioration of soil fertility. The second group is premised on the incompleteness of the reforms; underdeveloped land legislation and problems with establishing family farms are among the shortcomings cited. Thus, dissatisfaction with land reform appears on both sides—among those who strongly favor the reforms and those who oppose them.

Farm Restructuring

Although more than three quarters of the experts saw land reform as crucial for Russia, only 60 percent accepted the need for farm restructuring. Again, farm managers demonstrated more conservative views, with only 40 percent supporting restructuring. Naturally, the political left resisted restructuring, and the right was more supportive; still, 33 percent of the left and 55 percent of the center left agreed that farms should be reorganized. Age did not much influence the opinions of the experts.

In contrast to their responses on land reform, only 62 percent of the experts thought that farm restructuring certainly (23 percent) or probably (29 percent) had not been carried out. Farm managers strongly inclined toward this view (73 percent). Representatives of the mass media were even less optimistic; only 25 percent said restructuring was taking place. Officials were relatively optimistic by comparison, with more than 40 percent convinced that farm restructuring had been implemented. The political right was more likely to say that restructuring had occurred (44–49 percent), an opinion not shared by many of those on the left and by only 19 percent of the extreme left. Almost half of younger experts, but only a quarter of the oldest group, expressed optimism about the implementation of restructuring. In all, 15 percent of the respondents thought restructuring had been carried out, and 23 percent thought it probably had.

As with land reform, Pskov experts leaned to the view that farm restructuring had been carried out (50 percent). The most negative regional responses were those of Moscow (30 percent) and Irkutsk (26 percent).

In contrast to the responses on land reform, less than one third of those who thought farm restructuring had been implemented agreed with the underlying concept. The categories most supportive of the concept of farm restructuring (more than 50 percent of those who thought it had been implemented) were agribusiness managers, the mass media, the political right, and the under-35 age group. Only 12 percent of farm managers who responded positively about implementation supported the concept.

Among the regions, Tomsk is striking, with its extremely low proportion (9 percent) of agreement with the concept of farm restructuring. Other regions demonstrate a similar distribution of the responses to the average.

The general evaluation of farm restructuring was the same as for land reform: a score of 2 points on the scale from 1 to 5. The political right gave it a somewhat higher 2.3, and the left scored it at 1.3–1.6. As was consistent with their low acceptance of the farm restructuring concept, Tomsk experts gave the lowest mark for implementation (1.7).

Thus, the farm restructuring concept received less support from the agricultural establishment than did the land reform. As the land reform concept was closely connected with the reorganization of the *kolkhozy* (collective farms) and *sovkhozy* (state farms) and with the concept of land and asset sharing, the survey results may mean that the opponents of farm restructuring consider land reform necessary for Russia but oppose distributing land to households. It may also mean that they support the transfer of landownership to the large-scale farms as legal entities and disagree with the allotment of land shares to individuals.

Such a conclusion is corroborated by the debates on the new Russian Land Code in the legislature and in the society. The Agrarian Faction in the State Duma (the lower house of the parliament) and one faction of the Agrarian Party advocated fixing landownership rights in the large farms and not giving members the right to withdraw from the cooperatives with their plots. The reformers backed the 1991 concept of land reform, under which individuals would receive property rights in land shares. That concept has implications for the restructuring and functioning of large farms; for example, farm managers would have to rent or buy shares from individual shareholders, which would be unnecessary under the Agrarian Party's formulation. The upshot is that the level of disagreement regarding farm restructuring is even higher than in the case of land reform.

The respondents mentioned, among the positive consequences of farm restructuring, the creation of private, independent production units that own property and can choose the legal form of their business. Negative evaluations cited the formalistic approach to restructuring, the destruction of the technical potential of agriculture, the disruption of rural life, and the incomplete implementation of the concept.

Privatization of the Food Industry

The necessity of privatizing the food industry was even less obvious for the experts than the need for land reform and farm restructuring. Only about 52 percent thought that this measure was really needed. Even among agribusiness managers, only 57 percent believed food industry privatization to be crucial, and almost two thirds of farm managers opposed it. The under-35 group and the political right support privatization; the oldest group and the left are more likely to oppose it.

The regional distribution of opinion on the necessity of food industry privatization is very high in comparison with other components of the reforms: it ranges from 38–40 percent in support in Orel and Pskov to 70–72 percent in Voronezh and Irkutsk.

More than 70 percent of the experts did not agree with the concept of privatization in the food industry sector. (About half of this group was in total disagreement; the other half "disagreed more than agreed.") Among farm managers and academics, the share of nonsupporters was 82 and 84 percent, respectively, and for agribusiness managers, the figure was 67 percent. Of the younger experts, 41 percent felt privatization was necessary, but only 12 percent of the oldest group did. Support for food industry privatization on the political spectrum falls sharply from 36 percent on the right to 15 percent on the left. Among the regions, Orel and Moscow were least supportive of the concept (14 and 11 percent, respectively).

Despite the high level of disagreement with the concept, the evaluation of its results, averaging 2.4, was higher than for the reform components discussed above. The political right gave the reform a score of 2.6–2.8, the left, a score of 1.6–1.9. Among the regions, Orel experts recorded the lowest evaluation (1.5); the other regions were relatively even in their rating of food industry privatization.

In the experts' comments, positive consequences cited included growth of foreign direct investment, changes in the product mix, and increased compatibility of the industry and the market. The main negative consequence mentioned was monopsony. The agricultural establishment is extremely concerned about local monopsonies of processors, which depress farmgate prices and contribute to farm insolvency.[1] These attitudes toward processors lead experts to oppose food industry

---

[1] Studies show that these attitudes toward the processing sector are not completely warranted (see, for example, Serova and Melukhina 1995). However, the survey indicates that anxiety about monopsony is widespread in the society.

privatization in principle in order to avoid real or putative monopsony. That Irkutsk experts tend to agree on the need for privatization is noteworthy. In that large oblast, farmers would find it difficult to deliver to any processing plant but the local one because of the distances involved. Irkutsk is thus more likely to face local monopsony in the processing industry than Orel or Pskov, yet experts from those regions voice more objections to privatization than do those from Irkutsk.

Development of Market Infrastructure

Table 4 presents the experts' evaluation of elements of market infrastructure, on an ascending scale of 1 to t. They gave a rather high score to the emergence of organized markets for agricultural and food commodities (2.5) and of a network of intermediaries (2.8). These systems were to be set up at the initial stage of the reforms, as the distribution network that served the centrally planned economy was not adequate for a market environment. Two other elements of the agricultural infrastructure—education and agricultural research—remain almost unreformed. Nevertheless, respondents gave them relatively high scores: 2.7 and 2.8, respectively. More surprisingly, the standards and quality control system in the agricultural and food chain received about the same score (2.6), which would seem to be an overestimate. The explanation may be that the experts are not fully aware of what these systems should be like in a market economy and consider the institutions inherited from the Soviet economy appropriate for the new economic paradigm. Agricultural cooperatives and producers' unions, taxation, input supply, and market information, were valued rather modestly (2.1–2.3). As expected, bankruptcy legislation, antitrust regulation, crop insurance, and agricultural credit received low scores (1.8–2.0).

The overall evaluation for all infrastructure elements is 2.7, which is higher than the scores for any of the reform components discussed earlier. It is especially interesting that in this particular area there was no explicit government policy to facilitate the emergence of infrastructure.

The highest scores for infrastructure development were given by officials, the younger experts, and the political right (2.9–3.0). The lowest scores were given by academics (2.4) and by the oldest group and the left (2.5 for both).

It is notable that development of markets receives a relatively high rating from all experts except farm managers. This result testifies to the adequate level of development of food markets and the persisting underdevelopment of wholesale agricultural markets.

Table 4. Evaluation of Elements of Market Infrastructure

(score, on a scale of 1, lowest, to 5, highest)

Antitrust regulation	1.99
Bankruptcy legislation	1.92
Science and research	2.79
Education and Training	2.708
Cooperatives	2.25
Producers' unions	2.175
Food standards and safety controls	2.58
Market information	2.28
Crop insurance	1.82
Taxation	2.32
Agricultural credit	2.32
Input supply	2.067
Intermediaries	2.84
Organized markets	2.52

Agricultural and Food Policy

The average score for the agricultural and food policy of recent years was 2.1. All of the groups concurred in the relatively low evaluation of current policy. The dispersion of the estimates, at 0.684, is the lowest of among the scores for the components of the reform

As is shown in Table 5, the three main shortcomings of agricultural and food policy mentioned, in rank order, are the lack of budget expenditures for agriculture (65 percent of the responses), the inefficient utilization of agricultural outlays (41 percent), and the lack of a sound administrative system (35 percent).[2] The remaining reasons given for policy inefficiency lag well behind the first three. It is worth noting that the proponents of the less liberal policy model dominate the sample. Thus, overregulation of the sector was mentioned by only 3 percent of the responses, while underregulation was cited in 13 percent of the responses; high trade protectionism, in 5 percent; and low trade protectionism, in 11 percent. It is also remarkable in this respect that although an absolute majority of the experts considered inadequate budgetary allotments for agriculture the most important problem, less than half

---

[2] More than one choice could be recorded.

mentioned inefficient use of the allocated means as a shortcoming. Excessive regionalization (or centralization) of agricultural and food policy is not considered an important problem; only 4 percent of the sample was concerned about either.

Table 5. Responses on the Main Reasons for Inefficiency in Agricultural and Food Policy

Reason	Percentage of responses
Low trade protectionism	106
High trade protectionism	5.4
Excessive regulation	4.4
Excessive centralization	4.2
Inefficient administrative system	34.9
Underregulation	13.4
Overregulation	3.2
Inefficient use of budget outlays	41.0
Inadequate budget expenditures	64.7
Other	6.0

All groups of experts mentioned the top three reasons for policy inefficiency listed above, and almost all ranked them in the same order. The only exceptions were officials, agribusiness managers, and the political left, who ranked inefficient use of budget outlays third. In an implicit way, this substantiates the widespread opinion that the agricultural budget is spent in favor of downstream sectors rather than agriculture and that funds often leak to the left parties' political purposes. It is also interesting that both the under-35 age group and the political right, which on other questions demonstrated fairly liberal approaches, advocated more regulation of the agricultural and food sector and higher trade protectionism.

The more conservative segment of the sample (the political left and the oldest group), as well as the officials, were more likely to say that current policy is too highly concentrated in the regions, while the other groups inclined toward increased regionalization. Interestingly, regional officials did not consider the problem of policy regionalization at all important. (Only one respondent in this group considered the problem of excessive regionalization serious.)

In spite of the small deviations in the evaluations, a prevailing attitude toward current agricultural and food policy is evident: all groups in the agricultural establishment are dissatisfied with the policy and see as the main reason for its unsoundness the low level of budgetary financing of the sector. The absolute majority of experts tends to favor a less liberal policy than is implemented today.

General Evaluation of the Reforms

On average, all experts gave the completion of agrarian reforms in Russia a score of 2.2 (out of a possible 5) and the effectiveness of these reforms, a score of 2. The ratings were approximately the same irrespective of professional group, political alignment, or age. The political right (but not the center right) gave the reforms a slightly higher score, 2.3 for implementation and 2.2 for efficiency. Evaluations somewhat lower than average were recorded by academics (1.9 and 1.7) and by the oldest group (1.8 and 1.6). By region, the average estimates coincide with those of the entire sample. Only the Pskov experts gave implementation of the reforms a relatively high score, 2.4.

Political and Social Aspects of the Agrarian Reform

The experts were asked several questions regarding the social consequences of the reform in rural areas, political representation of the interests of the rural population, and the trend in public consensus on the reforms. Although these responses are not expected to measure the objective situation, they provide some indication of public attitudes.

Social Consequences

Almost 60 percent of the experts agreed that the reforms had led to a severe deterioration of living standards in rural areas, and nearly 20 percent believed that they had worsened living standards to some extent (Table 6). A minority said that the reforms had improved the condition of the rural population, somewhat or significantly. Almost 20 percent noted that a polarization in the living standards of rural residents had occurred and that part of the rural population had benefited from the reforms, while others had lost out.

The most pessimistic estimates were given by farm managers, 84 percent of whom believed that living standards in the countryside had fallen (or somewhat worsened) as a result of the reforms and 16 of whom pointed to a polarization in living standards. Those most optimistic about the trend in living standards were the under-35 group (11 percent positive responses), the political right (9 percent), and agribusiness (8.5 percent).

Table 6. Responses on How the Reforms Affected Social Conditions in Rural Areas

Response	Percentage of responses
Significantly increased living standards	27
Increased living standards somewhat	5
Decreased living standards somewhat	17
Sharply decreased living standards	58
Led to social differentiation	18

To sum up, despite some differences, supporters and opponents of the reforms agreed that the changes had led to deterioration of life in the countryside. Those respondents who saw such a deterioration rated the efficiency of the reforms at 2 (that is, at the average level). Those who believed rural living standards had improved gave the reforms a score of 2.5 on efficiency. Thus, in appraising the results of the reforms, the experts considered the social consequences as well as the economic achievements.

The Position of the Agrarian Party

The Agrarian Party is considered the main representative of the economic and social interests of the rural population in Russia (see Yanbykh, in this volume). In the survey the experts were asked to estimate the party's real role in representing rural interests. About two thirds of the experts were convinced that the Agrarian Party does not (28 percent) or probably does not (37 percent) adequately represent the rural and agricultural population.

Farm managers and academics—presumably the main contingent of the Agrarian Party—were a little less likely than the survey average to say that the party does not represent the interests of the rural population (58 percent of responses). Less than 50 percent of the political left thought the party was not representative, and the oldest group was more likely to take a positive view of the party. The extreme right—traditional opponents of the Agrarian Party—and the mass media were more likely to say that the party reflects the interests of rural people (48 percent negative responses). It is notable that the center right was the most pessimistic regarding the coincidence of interests of rural people and the Agrarian Party (66 percent negative responses).

Despite some deviations in the evaluations, it is obvious that the greatest part of the agrarian establishment does not see the Agrarian Party as a real political force championing the interests of agriculture and the rural population. The failure of the reforms and the worsening of living standards in rural areas are not brought into the political arena in an effective way. This may be a significant reason for the shortcomings of agrarian reform in Russia.

Public Consensus

More than the half of the experts (55 percent) believed that the dissensions within society regarding agrarian reforms have eased for the time being, while 28 percent said that the level of disagreement has not changed. A rather positive result is that only 13 percent of the experts observed a growth in the gap between approaches to the reforms. The lowest share of responses indicating strengthening of the contradictions was found among the mass media (10 percent) and the political left (11 percent). The left also showed the greatest confidence that social accord regarding the reforms was increasing (75 percent of the responses). An explanation may be that the radical stage of the reforms has already passed and that the conservative part of society (although not the oldest experts) has adjusted to and partly absorbed the reform ideas. Journalists simply register the trend. The political right, agribusiness, and academics are more likely than the sample as a whole to feel the lack of consensus; about 17 percent of them said that differences in approaches to the reforms are growing.

In summarizing this part of the survey, one can say that the probability of a consensus on agrarian reforms is greater today than at the start of the reforms. However, this consensus is more to the liking of the conservatives. The real transformations were still pending in the second half of the 1990s, and conservatives could be delighted with that situation. The political left has accepted the reforms in part, and the incomplete transition looks more acceptable to them than at the beginning. The more radical part of the society wants more revolutionary changes and cannot view the situation as moving toward consensus.

Prospects for Russia's Agriculture

Exploring experts' opinions regarding prospects for Russian agriculture was not the main objective of our study. However, the survey did include three questions related to this subject: What form of agricultural producer will prevail in the next decades? When will Russian agriculture start to recover? Can Russia become a net grain exporter, and if so, when? The distribution of the responses to these questions may also be indicative of public attitudes toward the reforms.

The Future Form of Agricultural Producer

Responses on the future form of production were mainly split between two possibilities: production cooperatives (48 percent) and large commercial companies (30 percent). Only 12 percent saw family farms as the prevailing type of production unit in the future. The experts had relatively high hopes for production on the subsidiary plots of rural households (26 percent).[3]

Farm managers saw production cooperatives as the most probable future form (55 percent of responses), as did the oldest experts (54 percent) and the political left (64 percent). The younger experts and the agribusiness managers were more likely to see companies as the main production units in the near future.

Among farm and agribusiness managers, the political left, and the young, few (16–18.5 percent) saw the household plot as a prevailing form of production. By contrast, the figure for mass media representatives was about 30 percent.

The share of experts who said family farms would become an important form of production ranged from 25 percent on the right to 4 percent on the left. The oldest experts (5 percent), officials (8 percent), and farm managers (8.5 percent) were unlikely to mention this sector.

The regional breakdown is notable. In Pskov, a marginal agricultural area, the experts most often foresaw the development of production cooperatives and, in second place, family farms. In Voronezh, a major agricultural area, the experts were less optimistic than the sample as a whole regarding the fate of the cooperatives but more optimistic about private (family) farming. Only 4 percent of Moscow experts (mostly federal officials and academics) saw private farming as the prevailing form. For Orel experts the share was only 3 percent of responses, although Orel was one of the first two regions to initiate private farming in the *perestroika* period.[4]

---

[3] More than one choice could be recorded.

[4] Orel's experiment with family farming was based on a faulty concept whereby farmers received all their initial capital from the administration (though into leasing). In Pytalovo *rayon* (district), Pskov oblast, another experimental region in the perestroika period, the farmers were provided with land, and the others had to earn their shares. Our survey demonstrated that in the experts' opinion, the family farm sector has better prospects in Pskov than in Orel.

Prospects for Agricultural Recovery

An absolute majority of experts (57 percent) said that Russian agriculture would not begin to recover within the next 10 years. One third said recovery would begin 5 to 10 years in the future; 10 percent said 3 to 5 years; and less than 0.5 percent said 1 to 2 years (Table 7). The mass media representatives were the most optimistic: 1 percent of the responses predicted that recovery would start in 1 to 2 years, and only 20 percent placed the beginning of recovery 10 years in the future. The left and the oldest experts were most pessimistic, with 75 and 63 percent of responses, respectively, ascribing the beginning of recovery to the next decade.

Table 7. Responses on the Recovery of Russian Agriculture and Grain Exports

(percentage of respondents)

Question	1–2 years	2–5 years	5–10 years	>10 years	Never
When will agricultural recovery begin?	0	10	33	57	
When will Russia become a net grain exporter?	1	6	21	36	36

Among the regions, Pskov was the most pessimistic (70 percent of the sample predicted the start of recovery after 10 years) and Irkutsk the most optimistic (2.6 percent hoped to see the beginning of the growth in one to two years, and only 51 percent pushed the prospects for recovery to the next decade).

To test the consistency of the responses, the correlation between answers on this question and the general appraisal of the reforms was examined. Those who did not expect recovery to start within the decade gave the reforms lower overall scores than the average (2 for implementation, versus an average 2.2; 1.8 for efficiency, versus an average 2). Those who predicted recovery within one to five years scored the reforms 2.3 for implementation and 2.2 for efficiency. In other words, the experts associate their expectations of economic growth in agriculture with the progress of the agrarian reforms.

Prospects for Grain Exports

Historically, Russia was a great grain-exporting state. Climate conditions have not changed, and only the loss of Ukraine after the breakup of the U.S.S.R. makes any difference to export opportunities. One might expect that Russia would be able to reclaim its status as a net grain exporter.

The experts' estimates of Russia's grain-exporting opportunities, however, were even gloomier than their views of the prospects for recovery. More than one third of the respondents saw no chance that Russia would become a net exporter of grain; another one third said it would happen after 10 years; 6 percent expected it to occur within 5 years; and only 0.6 percent said it would happen within 1 to 2 years (see Table 7).

The mass media, academics, and agribusiness were more optimistic regarding grain exports, while officials and farm managers were more pessimistic. As usual the younger experts expressed more hopeful views than the older ones. It is interesting that on this question the extreme political wings agreed; neither the right nor the left saw near-term grain-exporting prospects. Respondents who identified themselves as center right or center left seemed less pessimistic.

The Siberian net-importing regions were less optimistic than others; almost half of the Tomsk sample was convinced that Russia will not be a significant grain exporter at all, and in Irkutsk 38 percent of the respondents expressed this opinion. Respondents in Rostov, a major grain-producing area, were the most hopeful regarding grain exports.

The consistency of the experts' responses was again tested. Those experts who saw no chance for Russia in the world grain market also rated the implementation and efficacy of reforms lower than average (2 for implementation and 1.8 for efficiency). Those who optimistically foresaw the beginning of serious grain exports within a five-year period evaluated the reforms higher than the average (2.3 for implementation and 2.2 for efficiency).

Conclusions

The survey demonstrated that the greatest part of the agrarian establishment admits the necessity of agrarian reforms in Russia. In particular, the need for reform of land tenure is accepted. However, an absolute majority is dissatisfied with the actual reform process. Despite eight years of debate and implementation, the view of the bulk of the experts is that the reforms have not been carried out, and among those who say they have been carried out, a large share does not accept the concept behind them. The general evaluation of the completion of the reforms and their efficiency is very low on average, and the variation in the estimates is insignificant (Table 8). Thus, the agricultural establishment is unhappy with the reform process in Russia agriculture. The dissatisfaction extends to all the groups that made up the sample, but sometimes for different reasons. The more radical part of the establishment (the young and the political right) wants to see deeper changes in the agricultural system, while the more conservative groups (older people and the political left) dislike what they view as the excessive modifications of the system.

Although most of the sampled experts approve of land reform, the implementation of this particular reform component garnered the lowest scores on evaluation. By contrast, experts are very concerned about the food industry privatization but give implementation of that component a high rating.

The agricultural establishment is rather dissatisfied with current agricultural and food policy. The respondents agree that the inadequacy of budgetary expenditures for the sector is the main problem in this area. There is almost no difference among professional, age, and political groups in their inclination toward a less liberal agricultural and food policy. Despite the widespread opinion that there is a conflict between regionalization and centralization in agricultural and food policy, the experts do not see any problems in this respect. If there is some insignificant difference in approaches, the balance of responses favors a tendency toward centralization.

In giving the reforms low evaluations, the experts considered the severe social consequences of the reforms for the rural population: an absolute majority of the agricultural establishment believes that living standards in the countryside have declined. Young respondents, the political right, and agribusiness managers, however, look at the problem more optimistically, and a significant part noted that the rural population has benefited from the reforms. Since the objective of our survey was to measure opinion, not actual conditions in the countryside, we cannot conclude from the results that living standards have fallen. However, the agreement of the experts on this point does demonstrate the real attitude of different parts of the agricultural establishment toward the reforms and their consequences.

The Agrarian Party presents itself as the principal representative of the interests of the rural population, but it is not so viewed by an absolute majority of the agricultural establishment. There seems to be room for further party building in the society in order to reflect in the political sphere the interests of agriculture and of rural dwellers. Another possibility would be the radical transformation of the existing Agrarian Party, and indeed this process is already under way to a degree, as the party begins to float to the right of the Communist Party of Russia.

Table 8. Evaluations of the Reforms and Their Components

Component	Score	Dispersion
Land reform	2.01	0.792
Farm restructuring	2.07	0.744
Privatization of the food industry	2.37	0.918
Agricultural and food policy	2.06	0.684
Implementation of the reforms	2.17	0.744
Efficiency of the reforms	2.00	0.648

Note: The score is on a scale of 1 to 5, with 5 the highest. The formula for dispersion is

$$\frac{n\sum x^2 - (\sum x)^2}{n(n-1)}.$$

In general, respondents noted an increasing social consensus regarding the agrarian reforms. The left is more optimistic on this point than the right. This result may mean that the conservative part of the society accepts the concept of transformation for the time being and that resistance to more profound transformations will ease sooner or later. Then it would be worthwhile to postpone radical changes for the sake of consensus, even though delay in completing the reforms significantly worsens the circumstances for their implementation under the prevailing conditions in Russia. However, this increased agreement, mentioned mainly by conservatives, may also mean that the reforms are stalled, to the satisfaction of the opponents of the changes. In that case, new, radical transformations will be possible only if there is fundamental change in the political balance of forces: thoroughgoing reforms cannot be launched without broad public support.

In accordance with their low evaluation of the reforms, the experts are generally pessimistic in their forecasts for Russian agriculture. The great bulk of the sample sees no possibility for agricultural recovery within the decade and does not foresee a Russian niche on the world grain market. Although the survey measures opinions and does not pretend to make forecasts of the real economic situation, the results do show the mood of the agricultural establishment, which inevitably affects behavior in business, policymaking, and political life. Lack of hope promotes ad hoc policies and decisions that maximize short-run benefits, leads to reluctance to make investments, and hampers normal political representation of the sector. Russia's agriculture needs some clearly positive (even if not very important) achievements to break up this extremely negative trend, otherwise, there is a danger of self-fulfilling prophecies.

It is notable that the agricultural establishment still believes that cooperatives will be the prevailing form of production unit in Russian agriculture in the future. The low deviation of the responses on this question has two implications. First, part of the sample of experts is still convinced that the cooperative form is the most appropriate for agriculture (at least, in Russia). Neither current economic theory (not well known in Russia, by the way) nor 10 years' experience has shaken this conviction. Second, the more advanced part of the establishment discounts the possibility of changes in the current form of production because of the stalled reforms. This attitude needs to be changed—through education, public information, and training, and by escaping from the danger of a self-fulfilling prediction through promotion of experiments that can serve as examples.

In spite of deviations in attitudes and estimates for certain groups (defined by profession, age, political conviction, and region), the general findings are universal for the entire agricultural establishment. That makes such attitudes a serious political force in the society, and one that cannot be ignored. Although the position of the agricultural establishment is not well represented on the political scene, policymaking should take its views and approaches into account for the sake of successful implementation.

References

Serova, Eu., and Melukhina. 1995. "On Monopolism in the Processing Industry." *Issues of Economics,* no. 1: 67–76.

Table A-1. Composition of the Expert Sample

Category	Number of persons
Professional group	
Federal legislature	33
Federal executive bodies	69
Regional officials	54
Agricultural managers	83
Agribusiness managers	69
Science, education, training, consulting	126
Mass media	126
Others	48
Age group	
35 and younger	125
36–50	227
Over 50	146
Gender	
Female	224
Male	276
Political group	
Right	61
Center right	140
Center left	131
Left	52
No political identification	118
Regional group	
Irkutsk	81
Moscow	46
Orel	92
Pskov	96
Rostov	13
St. Petersburg	2
Tomsk	101
Voronezh	71

Paradoxes of Agrarian Reform in Russia

Zemfira I. Kalugina

Russia has arrived at the end of the 20th century with many of its social problems, including that of food provision, still unresolved. Food production, processing, and distribution in the 1980s were not adequate to provide a full, balanced diet for the population. In a number of areas, food distribution was rationed (Kalugina 1991a: 168). Attempts to improve the situation via the administrative command system through superficial adjustments turned out to be futile in the long run. The underlying reason was that the social and economic innovations which were introduced, such as intrafarm cost-benefit analysis, various types of contract, and intensive technologies, did not get at the core of the problem. They could yield only short-lived improvements, and only for specially chosen, artificially created experimental farms that enjoyed more favorable conditions than elsewhere. After each campaign, everything returned to the original status. The socialist system repelled the market elements alien to it. If the situation was to be reversed, thorough reforms would be needed.

The radical economic transformation initiated at the beginning of the 1990s was aimed at making constructive changes in the national agrarian sector. It included land reform, reorganization of collective and state farms (the dominant forms of socialist agrarian enterprise), and the development of independent private farms.

The Law on Land Reform passed in December 1990 repudiated the state monopoly on land and reinstated private ownership of land—a right also secured by the Constitution of the Russian Federation. Yet in 1990, the same year that the land reform law was enacted, the second (extraordinary) Congress of People's Deputies instituted a 10-year suspension of land transactions. This suspension is still in force despite the decrees issued by the president of the Russian Federation aimed at protecting citizens' constitutional rights to land and canceling the suspension. Debate on liberalized land transactions continues.

At the end of December 1991 the Russian government introduced provisions concerning the reorganization of collective and state farms and the order of their privatization. These measures were aimed at changing the organizational and legal status of collective enterprises, giving workers the right to free choice of a form of entrepreneurship, and endowing workers with shares of assets and land, along with the right to leave the collective enterprise without having to ask the working collective's permission. Reorganization was to reach every collective enterprise, profitable or unprofitable. Various partnerships, joint-stock companies, agricultural cooperatives, independent farms, and associations could be set up. The working collectives were also allowed to retain, if desired, the parental form of economic activity. Reorganization was to be completed by the end of 1992.

The development of a private farm sector began with the adoption in December 1990 of the law "On Peasant Farms." That law laid down the economic, social, and legal basis for the organization and operation of private farms and farmers' associations.

By the early 1990s, then, the legislative basis for the creation of a mixed economy in the agrarian sector had been created, and every rural worker had a choice of forms of landownership and farm management.

Trends in the Agrarian Economy during the Reform Process

The reorganization of collective and state farms was practically complete by the beginning of 1994, with 95 percent of collective enterprises reregistered. Under the reorganization, 66 percent of collective agricultural enterprises changed their organizational and legal status, and 34 percent exercised their right to retain their parental form. After reorganization there were 300 open joint-stock companies, 11,500 partnerships of all types, 1,900 agricultural cooperatives, 400 subsidiary farms belonging to industrial and other institutions, 900 associations of independent farms, and 2,300 other forms of organization. About 3,600 state farms and 6,000 collective farms retained their existing status. Among agricultural enterprises 26.6 percent were state owned, 1.5 percent were owned by a municipality, 66.8 percent were privately owned, and 5.1 percent were under a mixed form of ownership (Goskomstat 1995: 48–49).

Collective Enterprises

The reorganization of collective enterprises was the first step toward the creation of a mixed agrarian economy. Unfortunately, the reorganization failed to achieve positive results such as increased efficiency and higher output. In fact, the agricultural output of collective enterprises and their share in output are declining steadily (see Table 1). Collective enterprises accounted for 74 percent of total output in 1990 but for only 52 percent in 1996 (see Figure 1). The number and the productivity of livestock owned by collectives continue to decrease. Most collective enterprises are in a precarious economic position. At the end of 1991 their average profit was 43 percent, reflecting their capitalization; in 1995 it was –2 percent and in 1996, –20.5 percent (Stroyev 1997: 343).

V. Khlystun, former minister of agriculture and food of the Russian Federation, explained the situation in Russian agriculture as stemming from several factors: the constantly increasing disparity between prices for agricultural products and for agricultural inputs; the extremely low state subsidies; low purchase prices; delays in settlement of accounts for products sold; and monopoly positions held by processing, procurement, and service enterprises and organizations (Khlystun 1997: 7).

Figure 1. Contribution of Different Types of Farm to Russia's Agricultural Output, 1970–96

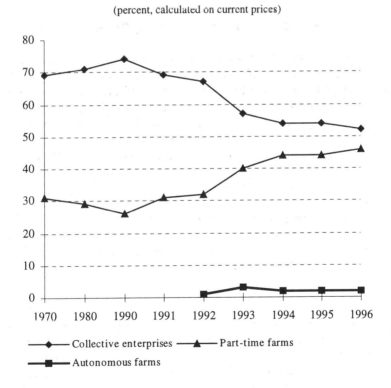

Source: Goskomstat (1997): 379.

Table 1. Agricultural Output in Russia, by Type of Farm

(comparable prices, percentage of the previous year)

Year	All types of farm	Collective enterprises	Household farms	Independent farms
1991	95	91	109	n.a.
1992	91	83	108	—
1993	96	91	103	167
1994	88	84	95	86
1995	92	85	103	97
1996	93	88	98	89

Source: Goskomstat (1997): 379.

In each Russian province some agricultural enterprises are functioning successfully despite the current unfavorable conditions. What is notable is that these farms have managed to adjust promptly to the new economic environment: they have studied the market situation, identified the most profitable channels for selling their products, and restructured their production according to market requirements. They have successfully developed processing of agricultural products and are selling their products through a network of their own stores, retail markets, or trusted wholesale agents, at better prices than they could otherwise command.

Some of these farms have set up large commercial structures and modern agroindustrial companies. They have the potential to become centers of scientific and technological progress in the countryside and to show other farms how to work under market conditions. But expert estimates put the share of such farms in Russia at no more than 5–7 percent (Poshkus 1997: 14).

Independent Farms

As a consequence of the adoption of the law on independent farms and the reorganization of collective and state farms, Russian peasants now have a real possibility of becoming independent economic agents. Between 1991 and 1997 the number of privately run farms grew to 279,000. In 1994, however, the growth rate began to decrease (Figure 2), and failures of independent farms are increasing. The number of bankruptcies was 5,100 in 1992, 19,100 in 1993, and 45,900 in 1994. In the second half of 1994 the number of bankruptcies for the first time exceeded the number of newly created farms (Goskomstat 1997: 379–83). Studies show that the main reasons for the instability of independent farms in Russia are extremely high taxes; exorbitant prices for agricultural equipment, fuel, and other resources; violations of owners' rights; low subsidies from the state; allotment of low-quality, remote land parcels; and lack of roads and communications (Petrikov 1995, 1996; Kurtsev 1996). The failures are also partly attributable to subjective causes such as Russian peasants' lack of experience with independent economic activity and low tolerance for economic and social risks.

Figure 2. Independent Farms in the Russian Federation, 1991–97

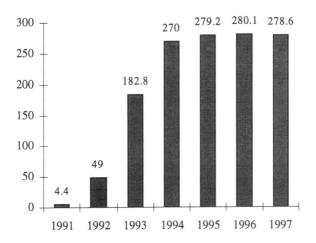

Number of independent farms (thousands)

As of January 1, 1997, the 279,000 independent farms occupied 12.2 million hectares, or 44 hectares per farm. Agricultural land in independent farms totaled 11.3 million hectares (93 percent of the allotted land); this included 8.3 million hectares of arable land (68 percent of the allotted land). Independent farms accounted for 5.3 percent of all agricultural land and 6.3 percent of all arable land. Over half of the independent farms had 20 or fewer hectares, and a fifth had 21 to 50 hectares. No more than 9 percent had over 100 hectares (Agriculture in Russia 1997: 3–10). The proportion of independent farms in total agricultural output is consistently low and does not exceed 2 percent.

Figure 3. Area of Land Plots Allotted to Independent Farms

(hectares)

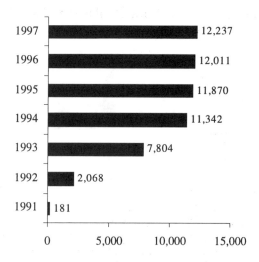

Source: Goskomstat (1997): 383.

In 1991, after analyzing the prerequisites and potential for the development of independent farms in Russia, we concluded that blanket decollectivization was too rash. Considering the state of public opinion, the level of industrial potential, the condition of the legislative base, and the social and political situation in the country, and taking into account that transition to a market economy would be a long process, we concluded that in the foreseeable future, the independent farm could not become the dominant form of agricultural production in the Russian countryside and that the most probable outlook was for the establishment of a mixed agrarian economy, with independent farms as one of the sectors (Kalugina 1991b). These predictions were fulfilled.

Household Farms

Household farming is a specific segment of the agrarian economy based on the resources and labor potential of rural families. Household farms appeared at the end of the 1920s in the process of socialization of individual peasant farms. The underlying principle was that although the state owned the means of production, including land, family members could work household farms, with no use of hired labor allowed. In 1991, under the Constitution of the Russian Federation, the household plots were transferred to citizen ownership. As a rule, the household farm is a sphere of secondary employment that supplements primary employment in the public sector of agriculture.

Figure 4. Structure of Agricultural Production of Russia by Type of Farm, 1990 and 1996

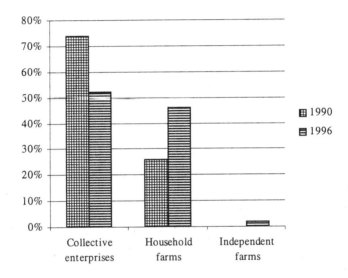

Source: Goskomstat (1997): 379.

Russian law now recognizes household farms as a form of agrarian enterprise, with full rights. All constraints on the number of animals that can be held by a household have been removed. Household plots can be enlarged up to 1 hectare by redistributing land under the management of local councils. In addition, rural inhabitants who were granted land shares (agricultural workers, pensioners, and some categories of workers in the social sphere) are entitled to use their shares to extend their household farms.

In the context of agrarian transformation, the partial disintegration of collective farming, and the lag in the emergence of new economic forms, the importance of household farming as a flexible, relatively stable, and self-regulating form of agricultural organization has increased. Whereas agricultural output in the public sector has declined, production by household farms has been increasing; in 1996 it made up 46 percent of total national output. At present, in a third of Russia's regions, households produce over 50 percent of agricultural products (Agriculture in Russia 1997).

Household landholdings fall into the categories shown in Table 2. A total of 6.1 million households have cattle, 4.1 million keep pigs, and 3.0 million keep sheep and goats. On average, per 100 households there are 75 head of cattle, 46 pigs, and 100 sheep and goats (Goskomstat 1995).

Table 2. Household Producing Units

Type of holding	Area (millions of hectares)	Number of households (millions)	Average size of holding (hectares)
Household plots	6.1	16.6	0.37
Plots in collective gardens	1.2	14.8	0.08
Collective kitchen gardens	0.6	7.6	0.082

In 1995 agricultural production by household farms stabilized or even decreased. During January–August 1995 households produced 1.9 million tons of meat (98 percent of the output during the same period in 1994), 12.1 million tons of milk (99 percent), and 7.5 billion eggs (90 percent). In 1996 the numbers of different kinds of livestock fell by 2 to 8 percent (Agriculture in Russia 1997; Goskomstat 1995).

There are several reasons, in our view, for the emerging trend toward a decline in households' output.

1. The destruction of the productive potential of collective enterprises, which provide to household farms, gratis or on easy terms, such resources as fodder, seeds, young animals, agricultural equipment, and transport vehicles

2. The substantial decrease in the financial opportunities of rural families as a result of the decline in the standard of living, including depreciation of savings

3. The near-exhaustion of rural families' labor potential. Analysis of time budgets of the rural population shows that many families are running their household farms at the limits of their physical strength (Artemov 1995).

Under current law rural households may choose to operate their household plots in cooperation with other businesses and with assistance from collective enterprises or to transform their household farms into independent farms. About 20 percent of rural respondents polled in a survey (by Kalugina, Institute of Economics and Industrial Engineering, 1997) were aware of the availability of these options. An initial basis for conversion could be large, commodity-producing household farms; about a quarter of rural families are inclined in this direction. But over half of the surveyed rural families think it is impossible to make the change because they believe that household farms cannot do without assistance from collective farms.

Experts (566 professional and administrative workers in the agricultural sector) are more pessimistic about the prospects for transformation. No more than 2.9 percent of them believe that household farms can be viewed as a transition to the independent private farm, and 61.2 percent believe that household farms can be successfully developed only in cooperation with collective enterprises.

In our view, the existing tax system promotes the perpetuation of the current status of household farms. Household plots are practically exempt from income tax, and the small land tax levied on them does not affect their profitability—which is further substantially increased by access to the resources of collective enterprises. Most rural people understand and take advantage of the favorable position of the household farm, and they can foresee that a shift of household farms from the nonformal to the formal sector of the economy would lead to heavy taxes and to cessation of aid from collective enterprises.

The role of household farms in the process of establishing private farms is two-sided. Because they are being developed in parallel with and substantially at the expense of collective enterprises, the household farms help perpetuate and preserve the old system of economic relations. But they do foster skills in the frugal and efficient management of land and promote the social qualities and business attitudes required in a market economy. The characteristic features of operators of household farms and their families are freedom of activity, independence in economic decisionmaking, and full economic responsibility for the results of their work. In other words, household plots help shape economic agents of a new type.

At present, most rural inhabitants do not dare undertake operation of an independent farm. Current realities, however—the decline of production in the public sector, very low wages in agriculture, irregular payments, and increasing unemployment—are forcing them to expand the scale of their household farms and improve the quality of the commodities produced. In size and function, these farms are becoming similar to independent farms, and former collective and state farmers are almost involuntarily becoming independent farmers.

The latent processes going on in the contemporary Russian countryside have so far remained out of public view and need to be thoroughly researched. It is these processes that can mold the trends and character of the changes in the Russian countryside in the near and middle term.

Outcome of the Reforms

Judged by formal indicators, the planned transformations have achieved some of their purposes: the number of collective and state farms has been reduced, and signs of a mixed economy and of diverse forms of ownership have appeared. But what is the merit of these transformations, and what is their social price?

Illustrations of the costs are not hard to find. According to estimates by the Russian Academy of Agricultural Sciences the number of cattle in Russia has declined to the level of over a quarter-century ago; animal productivity is also at about that level, and technical equipment is at the level of almost a half-century ago Over the period of implementation of the reforms, the standard of living of the urban and rural population, including consumption of staple foods, has declined sharply. In comparison with 1985, per capita consumption of meat and meat products has fallen 15 percent; dairy products and eggs, about 20 percent; sugar and vegetables, more than 30 percent; oil, 40 percent; and fish, about 60 percent. The only increase observed was in consumption of potatoes and baked goods (Agriculture in Russia 1997). Russia slipped from the 7th to the 40th position among industrial nations in provision of food products. According to Goskomstat, in 1995, 54 percent of the food consumed by households was imported, using external loans for payment. In 1995 the average citizen consumed an estimated 2,300 calories per day, although (taking into account climate conditions) a minimum of 3,200 calories was needed. Malnutrition, among other factors, affected health status, the death rate, and life expectancy (which dropped from 68 in 1990 to 62 in 1996).[1]

---

[1] Stroyev (1997); for life expectancy, see Rutskoy and Radugin (1997).

Examining the Paradoxes

The dynamics of the development of the three segments of the agrarian economy clearly shows the first paradox of the agrarian reform: the *expansion of small-scale commodity production.* Contrary to the reformers' intentions, the leading sectors in agricultural production are currently not independent farms or joint-stock companies but rural dwellers' household farms. The operators of these farms have no means of mechanization, partly because of the tight market for small equipment suitable for use on private plots but mainly because of lack of resources for purchasing equipment. What has enabled the household farms to double their output and become more commercial is the labor inputs of their owners and family members. But the expansion of small commodity production has many shortcomings: (a) this form of production becomes institutionalized; (b) there is a return to trade by barter; (c) the technological level of production declines; (d) the requirements of agrarian technology are not met; and (e) environmental problems appear.

The second paradox is the *inefficiency of "capitalization" in the agrarian economy.* The policymakers themselves have had to admit that in place of the inefficient state sector, there is now an inefficient private sector. In our view, the underlying cause is the formal character of the reforms: the organizational and legal status of collective and state farms was changed, but economic relations remained, in essence, the same. The position of workers in the system of production relations has hardly altered.

As most workers have never felt any difference between their previous status as employed workers and their present status as co-owners, no distinct change has occurred in their work motivation or behavioral patterns. The economic mechanism by which peasants can exercise their rights of ownership to their shares of land and other assets is not yet in operation. As was found in our surveys in Novosibirsk oblast, over 80 percent of the respondents had received no dividends on the shares of assets and land that they had handed over to agricultural enterprises for use. Most enterprises are pinched and unable to pay dividends to their workers. This situation is typical of other regions as well (Orlov and Uvarov 1997).

The third paradox of the reforms is that instead of developing in people a market mentality and market-oriented behavior, they are, in fact, contributing to the *destruction of motivations and rewards for work.* This is true in the economy in general, but the most vivid consequences are seen in the agrarian sphere, with its gap between workers' orientation toward higher earnings and the declining ability of agricultural enterprises to reward their contribution. Wages of agricultural workers are the lowest in the country—less than 40 percent of the national average and below subsistence level. Moreover, wages are systematically delayed for months.

Table 3. Provision of Social Services in Rural Russia

Type of facility	Annual average, 1991–95	Percentage of 1986–90 level	1996 (estimate)	Percentage of 1991–95 level
Residences (total floor space, millions of square meters)	10.3	53.6	7.8	75.7
Preschool institutions (thousands of places)	29.0	25.2	6.0	20.7
Comprehensive schools (thousands of places)	94.7	42.7	54.0	57.0
Clubs (thousands of places)	38.7	30.0	10.6	27.4
Hospitals (thousands of beds)	2.5	39.7	1.1	44.0
Dispensaries and outpatient institutions (thousands of visits per shift)	8.2	55.0	3.1	37.8

Source: Stroyev (1997): 343.

The connection between wages, work results, and worker's qualifications has been destroyed. One third of rural workers said that the size of their remuneration was not contingent on the enterprise's efficiency. Employment in the public sector has ceased to be a primary source of livelihood for rural workers. According to the data from our 1997 survey of the rural population in Novosibirsk oblast (N = 553), only 38 percent of respondents said that their primary source of income (monetary and in kind) was their wages; 42 percent said it was the household farm.

The ability of agricultural enterprises to solve their workers' social problems with their own resources has drastically decreased. Before reform, distribution of many social benefits was connected to the job. A worker could receive from his enterprise free housing, child care, medical treatment at health resorts, and so on. After reorganization, the collective and state farms were allowed to shift social and cultural services to local government—which, however, lacked sufficient financial resources or an appropriate material and technological base.

This led to a substantial worsening of social services in the countryside.

These processes are at the core of a sharply decreased motivation for professional, high-quality, effective work and a drastic drop in the prestige of work in the public sector, especially among rural youths. According to our 1997 survey, 31.9 percent of rural inhabitants would be inclined not to work at all if unemployment relief could provide a fairly good living. However only three years previously (in 1993, N = 525), no more than 10.6 percent of rural respondents held similar views. We see this increase as an alarming symptom.

The destruction of the system of compensation for work is reflected in attitudes toward work. People gradually lose their self-confidence and get used to state paternalism. Well over half (65.8 percent) of respondents preferred to have a guaranteed, even if small, income. Only 28.9 percent would be willing to accept high risk in exchange for high earnings. The emerging institutional context and the operating economic mechanism work to distort the system of individual values, decrease or negate the value of work in the public sector of agrarian production, promote "social infantilism" rather than develop market behavior and thinking, strip the transformations in the agrarian sector of their social base, and hinder modernization.

The fourth paradox is that the changes in the agrarian sector have brought with them the utter *poverty of the rural population* and the *degradation of the rural social sphere*. The cause, to a great extent, was the transfer of social infrastructure and services from the books of the agricultural enterprises to those of local councils, which lack the resources to maintain and develop social and consumer facilities (Table 3). As V. S. Chernomyrdin remarked, "[the councils] tried their best but got the usual results."

The social price that has had to be paid for the reforms has led to disappointment and loss of confidence in the wisdom of the changes. The consequence is increasing nostalgia for the past and for socialism. In recent sociological surveys, rural residents were asked to what degree their expectations of improvements as a result of the reforms had been fulfilled. For over 60 percent of respondents, the answer was, not at all; 20 percent experienced partial fulfillment, and only 10 percent said their expectations had been completely fulfilled.

Underlying Causes and Possible Ways out of the Impasse

The model of agrarian relations imposed from the top failed to take into account the traditions and historical experience of agricultural production in Russia or the joint collective-individual character of the development of agrarian relations.

For example, the reforms were drawn up without regard for the values of the rural population, most of which continues to be oriented toward collective forms of economic activity, corporate solidarity, and state ownership. As a consequence, the bulk of households did not embrace the principal ideas of the reform: liberation of business, private ownership of land, free sale and purchase of land, reorganization of collective enterprises, and development of independent farms. Professional and administrative workers in the agricultural sector withheld their support for rural transformation. A reform alien to the people and not supported by leaders is doomed to failure.

The reformers missed their chance to make maximum use of the social potential that had been present before the transformation. The high social price people have had to pay for the reforms has brought disappointment and doubt as to their wisdom. Outcomes for agricultural enterprises show that very few people can be successful under present conditions. Reform of the agrarian sector was carried out according to traditional Soviet "technology," with its typical features of command orientation, totality, enforcement, and tokenism.

Directors are unprepared for the new economic conditions. Yet it is the directors who in the future should be playing the main role in the transformation of the Russian countryside. Economic managers are demonstrating two types of economic behavior: conservative waiting and innovative activity. Very often, people who on a theoretical level vehemently oppose the reforms demonstrate standard market behavior in practice, while the supposed "champions of reform" turn out to be mere onlookers. The state structures themselves have proved unprepared for the extreme scale and speed of the reforms. This is evident in the lack of coherence in implementation of the reforms, the constant changes in the "rules of game," and the reckless elimination of state regulation of the agrarian-industrial complex.

The Russian model of agrarian relations should, in our view, draw on the prevalent system of people's values and take into account the great importance people attach to corporate solidarity. Even Western experts viewed as erroneous the Russian reformers' drive simply to root out people's "anticapitalist" mentality without being able to put in place a working substitute (Admiral Peter 1995). What is needed is a reasonable combination of collective forms with private initiatives, oriented toward balanced development of the three segments of the agrarian economy. Special attention should be paid to successfully functioning household farms, which should be given the opportunity to integrate into collective enterprises or to become independent.

The market mechanisms being formed should be accompanied by state regulation of the activities of agribusiness and associated branches, especially during the transition period. The rate, scale, and depth of the transformation should be calibrated with the development of the necessary social, economic, and legal institutions. Special attention should be paid to encouraging the development of new economic agents able to work under the emerging market-based conditions of economic and social risks. Finally, for a deep and objective assessment of the agrarian reform, it is necessary to provide for scientific monitoring of its progress nationwide and in individual regions.

References

Admiral Peter. 1995, "Where Is Russia Moving?" *Problems of Management Theory and Practice*, no. 4: 8–13.
Agriculture in Russia. 1997. "Agriculture in Russia in 1996: Economic Review by the RF Goskomstat." *AIC: Economics, Management*, no. 3: 3–10.
Artemov, V. A. 1995. "Changing Setting and Way of Life in Siberia (1972–1993)." *Sociological Studies*, no. 1: 73–83.
Goskomstat (State Statistical Commission of the Russian Federation). 1995. *Agriculture of Russia 1995*. Moscow.
———. 1997. Russian Statistical Yearbook 1997. Moscow.
Kalugina, Z. I. 1991a. Private Household Economy in the USSR: Social Regulators and Results of Development [in Russian]. Novosibirsk: Nauka.
———. 1991b: "Social Bounds to the Development of Autonomous Farms." *Izvestia SB AS of the USSR, Region: Economics and Sociology series*, issue 3: 35–42.
Khlystun, V. 1997. "To Stabilise the Operation of the Russian Agro-Industrial Complex." *AIC: Economics, Management*, no. 4: 3–16.
Kurtsev, I. V. 1996. *The Economics of Siberian Agro-Industrial Complex in the Period of Transition to Market*. Novosibirsk: Siberian Branch, Russian Academy of Agricultural Sciences.
Orlov, G. M., and V. I. Uvarov. 1997: "The Countryside and Russian Reforms." *Sociological Studies*, no. 5: 43–53.
Petrikov, A. V. 1995: "Specificity of Agriculture and Modern Agrarian Reform in Russia." In A. V. Petrikov, ed., *Encyclopaedia of Russian Villages*. Moscow.
———. 1996. "Agrarian Economics and Politics: History and Modern Years." In A. V. Petrikov, ed., *Encyclopaedia of Russian Villages*. Moscow.

Poshkus, B. 1997. "Inner Reserves of the Russian AIC." *AIC: Economics, Management,* no. 3: 11–19.

Rutskoy, A., and N. Radugin. 1997. "Agrarian Crisis Continues." *AIC: Economics, Management,* no. 1: 3–7.

Stroyev, Ye., ed. 1997. *The Conception of Russia's Agrarian Policy in 1997–2000.* Moscow: Open Company "Peak-Club."

Reform and Economic Behavior in Russian Agriculture

Eugenia Serova

The agricultural system that prevailed in the Soviet Union for six decades had a highly stable internal structure, but it also displayed deep-seated, intractable problems.[1] By the mid-1980s the Soviet type of large-scale farming was exhibiting growing inefficiency and was a burden on the national budget. Food shortages recurred from year to year. Numerous forced attempts to restructure the agrarian sector within the framework of the socialist economy had failed, and radical change had become unavoidable.

Even under the Soviet regime, the first moves toward agrarian reform were taking place. The collapse of the U.S.S.R. in late 1991 gave impetus to the radical stage of reform, which aimed at creating market-oriented production units. The free distribution of the land and nonland assets of *kolkhozy* (collective farms) and *sovkhozy* (state farms) among their employees and pensioners was the basic principle of Russian farm restructuring.

After almost a decade of restructuring, the initial objectives have not been achieved: most of the existing agricultural producers are not market-oriented units. The explanation for the modest results of the reforms is to be found in economic, legal and political, and psychological constraints.

Economic. Recessionary times do not create economic incentives for production units and therefore do not induce real transformation. For most farms, whether collective or individual, survival is the goal, and their activities are conducted accordingly.

Legal and political. Russia has no deep traditions of legal democracy and strict execution of legislation. Many laws are not actually implemented throughout the country. Contradictions and gaps in the reform legislation aggravate the problem. Lack of political consensus

---

[1] This chapter is based on the results of a study initiated and financed by the World Bank (Alex Norsworthy, project manager) and carried out by a research team consisting of Eu. Serova, N. Karlova, I. Kramova, S. Kramova, O. Pysmennaya, and T. Tikhonova.

within the society regarding the agrarian reform severely hampers the restructuring of the sector.

Psychological. The prejudices and the conservative mentality developed during the Soviet period retard the emergence of entrepreneurial activity in the countryside. Furthermore, the rural population is experiencing significant difficulties in adjusting to the new circumstances.

Nonetheless, despite all these constraints, the farming structure in Russia changed during the reform period in a number of ways. A completely new sector has emerged: that of individual farmers.

Farm restructuring led to a tremendous growth and strengthening of household production in both rural and urban areas. Large-scale enterprises remain the major agricultural producers in Russia, but they have become new production units operating in a completely new economic environment. As in all other transition economies, the farm market in Russia is weakly developed: less than 0.5 percent of farmland is involved in land transactions annually. However, the concept of farm restructuring and land privatization has led to the creation of an informal market in land shares.

These changes caused notable shifts in the structure of agricultural output, land use, and agricultural labor. Household production accounts for almost 50 percent of agricultural gross domestic product (agricultural GDP). After 70 years of state land monopoly, the state-owned share of farmland has decreased to one third; the remaining land belongs to individuals or to collectives of individuals. Employment in large-scale farms shrank by 45 percent during the reform period, the number of people involved in household agricultural production doubled, and a new form of agricultural employment—self-employed individual farmers and their hired workers—emerged. In addition to the structural changes, the reforms led to the emergence of new behavioral patterns in the agrarian sector.

Production

The lack of endogenous economic incentives for farms was for decades one of the main problems of Soviet agriculture. Profit did not play a role in state-owned agriculture because gross output, the structure of output, input and output prices, marketing channels, and wages were controlled by the state. Prices, which are a major market signal in a market economy, did not affect production (Table 1).

Table 1. Effect of Purchase Prices on the Economic Behavior of Agricultural Producers in the Prereform Period: 1991 Prices, Production, and Output

Crop or animal	Index of purchase prices (1988 = 100)	Index of sown area (1986–90 average = 100) or livestock headcount (1988 = 100)	Index of output (1986-90 = 100)
Grain	150	94	85
Sunflower seeds	104	105	93
Flax fiber	182	68	81
Sugar beet	113	95	73
Milk	118	97[a]	93
Cattle	146	94	94 (meat)
Hogs	160	91	
Sheep and goats	138	93	
Poultry	168	98	
Wool	151		91
Eggs	100		98

Note: Purchase prices were reformed in 1991. Input prices remained unchanged, so the higher prices for output, other things being equal, should have created incentives for producers.

a. Number of cows.

Party discipline exerted on managers was the main leverage for stimulating farm production. The most important indicator of farm activity was not output as such but fulfillment of the target figures, and managers strove to have lower targets assigned rather than try to produce more. The system led to false reporting and to overestimation of output. The 1980s saw many attempts to reduce the volume of state purchases and promote some form of quasi-market system, but these partial changes did not—and could not—solve the incentive problem in the framework of state agriculture.

In 1986, in an attempt to increase incentives for agriculture, the government allowed up to one third of total farm output to be sold on the open market. In the Soviet economy there was only one legal institution that could be considered an "open market": the town market (or, in the township, the kolkhoz market). These markets, however, did not have the capacity to handle the output of large-scale producers, and in 1988 only a few percent of output was sold outside the state procurement agencies. (The highest share, 6 percent, was recorded for fruit.) A survey of the best managers of that time showed that they did not see high revenues as the aim of their enterprises and did not consider marketing a necessary activity for farm management.

In 1991 the last centralized increase in purchase prices for agricultural products took place. Since input prices and the interest rate were not changed at that time, the rise in output prices should have led to increased production—if prices had been an incentive for the large-scale enterprises (kolkhozy and sovkhozy).[2] In fact, in spite of strict state control over production, the enterprises kept reducing the areas of those crops and the numbers of those animals that were not interesting from the point of view of the real incentives within the centrally planned economy.

The very first steps of the reforms altered the economic behavior of the farms in the direction of profit maximization. Farms started to choose the products that yielded the highest profit. Thus, 1992 already saw growth in the output of some cereals (such as buckwheat) that had been in permanent deficit in the Soviet period in spite of attempts to encourage their production by raising purchase prices.

Sunflower seeds are the most profitable agricultural product. They are a major item of Russian agricultural export and the only one that yields a positive trade balance. The area under this crop expanded in nearly every reform year despite the depletion of soils in the regions concerned.

From the first moment of liberalization, flax producers faced a poor market because of the cheap, better-quality textile imports flooding into Russia. Flax fiber output fell from an annual average of 124 million tons in the period 1986–90 to 54 million tons in 1994. The government announced a broad program of support for flax production, including state procurements, and in 1995 output jumped to 69 million tons. When the program failed for lack of financing, the area under flax and the yields again declined.

Although profit-cost ratios do not completely represent the economic incentives for agricultural producers, and their measures are not relevant in transition conditions, in Table 2 we attempt to show in a simplified way the reactions of agricultural producers to changes in these ratios.[3] Crop areas and numbers of animals (rather than output and

---

[2] The effect of the back-sloping supply curve, a phenomenon of small-scale family farming, could not occur in the Soviet economy, nor were there monopolies.

[3] "Profit" means the price of marketed product in relation to the production costs. In the Russian Federation, as had been true in the U.S.S.R., the profit-cost ratio is the primary indicator of the profitability of a product.

yields, which depend on the weather) are used to gauge producers' reactions to changes in the profit-cost ratio.

Table 2. Effect of Changes in the Profit-Cost Ratio on the Economic Behavior of Agricultural Producers in Russia

Crop or animal	1993	1994	1995	1996	1997
Year-to-year indexes					
Cereals (area)	0.98	0.92	0.97	0.98	1.00
Sunflower seeds (area)	1.01	1.07	1.32	0.94	0.93
Cattle (number)	0.94	0.89	0.91	0.89	0.90
Cows (number)	0.98	0.93	0.95	0.91	0.92
Hogs (number)	0.91	0.87	0.91	0.85	0.91
Profit-cost ratio (percent)					
Cereals	190	59	55	42	24
Sunflower seeds	—	145	134	30	18
Meat	64	-16	-20	-47	-55
Milk	8	-26	-1	-34	-33
Hogs	52	2	-4	-31	-31
Coefficient of correlation between indexes and the profit-cost ratio					
Cereals	—	—	x	—x	0.048
Sunflower seeds	—	—	x	x	0.781
Meat	—	—	x	x	0.950
Milk	—	—	x	x	0.438
Hogs	—	—	x	x	0.795

Source: Calculated from Goskomstat and Ministry of Agriculture and Food data.

Correlations between changes in production and market signals are positive and significant for the majority of products during the reform period. The insignificance of the coefficient for cereals may be an effect of the aggregation of all kinds of cereals.[4] During the period under consideration, feed and food grains showed opposite trends, which are not reflected in the aggregate figures. However, the changes in the profit-cost ratio for grain did affect the spatial structure of cereal production. The variation in profit-cost ratios for cereals by region in 1996 correlates with the variation of change in areas planted to cereals in these regions in 1997 (coefficient of correlation, $k = 0.686$). In other words, the higher profitability of grain in the region was reflected in an increase in the area sown to grain in the following year. That, too, indicates responsiveness to prices on the part of producers.

---

4. Disaggregated profit-cost ratios for different kinds of cereal are not available.

Marketing and Processing

Price liberalization and farm restructuring made it impossible to sustain the Soviet system of state purchases: gone were party discipline, state distribution of inputs, and, therefore, the leverage for enforcing mandatory deliveries by farms. Due to social inertia in the regions, managers and regional authorities still maintained some state purchases, but as early as 1993 the practice of deliveries to the state had started to disappear.

The network of former state procurement agencies was demonopolized and privatized in 1993. In 1996, following the experience of some Central and East European countries, the Russian government set up a procurement agency, the Federal Food Corporation. It soon proved ineffective (since it had no regional branches and no leverage over markets) and in 1998 was transformed into another unit, with regulatory rather than purchasing functions. The procurement system has changed radically: the state no longer has a monopoly on agricultural and food markets, new private intermediaries have been emerging, and a market is forming. The state presence has steadily diminished, and the share of state purchases has declined, especially for grains (Table 3).[5]

Not only were state purchases reduced; their nature had also changed. The government does not actually buy a certain volume of agricultural products but, rather, distributes soft budget loans to private procurement agencies in exchange for deliveries of products for state purposes (for instance, for army supplies). This field is one of the most corrupt in the Russian economy today. Under this system farms are not obliged to meet state purchasing targets.

With the end of the system of mandatory deliveries to state reserves, farms obtained freedom to choose buyers for their produce and so maximize their profits (or minimize losses). New institutions have begun to emerge.

---

[5] The share of state purchases of livestock products remains rather high, but that result is the effect of an erroneous statistical approach that counts transactions within regional boundaries as state purchases. The system was introduced in 1992 to reduce state subsidies for livestock products, which are paid only for deliveries to the state reserves. Several private (formerly state) processors are formally nominated as state purchasers, and all sales to those plants count as deliveries to the reserves and receive the subsidy. In accordance with the system, delivery of milk to a dairy plant in a neighboring region is not considered a state purchase, while delivery to a local plant is.

Table 3. Share of State Purchases in Total Sales of Selected Agricultural Products

(percent)

Crop or animal	1992	1993	1994	1995	1996[a]	1997[a]	1998[a]
Grain	64	63	33	28	26	25	8
Sugar beet	98	98	67	21	8	9	n.a.
Sunflower seeds	76	42	8	19	8	8	5
Potatoes	50	45	34	35	31	28	24
Cattle (live weight)	78	77	65	55	47	37	33
Pigs (live weight)	80	79	73	57	66	51	50
Poultry (live weight)	85	92	89	87	83	77	73
Milk and dairy products	96	96	93	90	84	81	80

a. Data are for deliveries to the purchasing organizations and are not necessarily the same as deliveries to the state. (Since 1996 Goskomstat has not registered purchases going to the state reserves.)
Source: Goskomstat data.

Commodity exchanges constituted the first experiment with this new approach, with cereals and white sugar being the major traded commodities. These exchanges took the form of spot transactions representing a very small segment of the total market. In some agricultural regions the commodity exchanges even had an effect on price setting. Since the fall of 1992 the commodity exchanges have been deteriorating because of state interference in prices and because of regional trade barriers.

The Ministry of Agriculture and Food made several attempts to create a network of *wholesale markets,* but in the event, such markets started to emerge spontaneously and in sites that were not officially planned. These markets serve small-scale farmers. Large farms, which remain major producers, tend to have direct contracts with buyers, so wholesale markets are not likely to play a significant role in Russia's marketing chains. They can be useful, however, for vegetables, fruit, and potatoes, mainly produced by households (which account for 70–90 percent of gross output.) It is exactly in these sectors that wholesale markets are now emerging.

Seasonal *agricultural fairs* have been revived. In periods of high inflation the type of contracting conducted at fairs would not be possible, but financial stabilization has made this institution useful again. (Sometimes the fairs are mistakenly termed wholesale markets.)

A completely new set of players has emerged in agricultural markets: *private intermediaries.* These can be huge, diversified companies or individuals operating at their own expense and risk. Rather large trading companies have been set up for the grain and oilseeds markets, and Russia's largest financial companies (Inkombank, Alfa-Kapital, MENATEP, and others) have branches dealing with purchases of these products.[6] In addition, there are new private traders such as the grain-trading company OGO.

Smaller intermediaries operate in the livestock market. In the dairy sector, firms trading in casein and milk powder are more or less widespread. Middlemen in the liquid milk market have scarcely developed, although as of 1998 there were signs of their emergence in major agricultural areas. In the meat market, small middlemen are rather common, especially in the purchase of meat from households. The largest meat plants in the country, faced with difficulties in the delivery of raw materials, have started to develop their own procurement networks.

The share of agriculture in final retail food prices was set too high in the Soviet economy, and price liberalization caused it to fall. To avoid a consequent drop in income, farms started to develop *on-farm processing facilities* ("small processing"). Except for bakeries, the efficiency of these enterprises turned out to be not very high. However, a certain part of agricultural output is processed on farms, and that determines the particular food distribution chain for meat, milk, and sunflower seeds.

Besides progressive trends in the creation of food distribution chains, there are negative trends, mostly connected with various *barter schemes.* Among the motives for barter are to compensate for the lack of working capital on farms and to avoid taxes. Agricultural producers use barter to obtain inputs, pay for energy, gasoline, lubricants, and fertilizers, and—a recent development—even pay taxes (which is definitely illegal). In addition, various types of in-kind payment to farm personnel for wages, dividends, and land rent became widespread in the reform years. Barter became even more common after the introduction of the government commodity loan program. In 1996–97 the amount of grain transferred through barter deals reached almost 70 percent of total sales.

One of the most widespread barter deals takes form of a trade credit: the supplier or sometimes the trader or processor provides inputs (fertilizers, fuel, and so on) to a farm in advance, before the sowing

---

6. It is not yet clear how the recent financial and economic crisis will affect the situation.

campaign, and receives the harvested crop. The terms of trade in such deals usually include implicit interest. Another type of barter consists of payment in output in return for electricity and other payables. In contrast to the trade credit, there are no preliminary agreements, and the terms are announced spontaneously. The third type of barter is the so-called *vzaimozachet;* in its simplest form, the regional authorities accept a farm's output and pay the farm's debt to the bank or other creditor—not necessarily in cash but with, for example, a tax concession.

The "barterization" of the agricultural and food economy makes markets nontransparent for players, which leads to price distortions. Since on agricultural markets the "lemon market" mostly favors buyers, the farms suffer greatly financially.

Nonetheless, the most advanced farms have assimilated the experience of working with private traders. They have begun to concentrate on marketing rather than on production—a tremendous positive shift after decades under a centrally planned economy.

The food industry was for decades a bottleneck in the Soviet agricultural and food sector, and numerous attempts to reorient state investments from agriculture itself to downstream sectors failed time and again. Besides, in the centrally planned economy, the network of processing plants could not have parallel units: there could be only one dairy plant in each *rayon* (district) and only one slaughterhouse for a group of districts. Coupled with the radial layout of road infrastructure, that led to the creation of potential local monopolies.

In the initial privatization program the food industry was envisaged as being part of the first phase of privatization. To avoid local monopolies, a special scheme was introduced for privatization of primary processors; farms had the right to buy controlling interests in the enterprises at nominal prices. In addition, regional authorities had long controlled prices and the margins of the processors, and a presidential decree envisaged regulation of managers' wages in the enterprises. As a result of all these measures, the processing industry, which could be an engine for the entire food chain, was in an even worse state than agriculture. Today these policies are not strictly implemented, and food processing has started to evolve in a direction more adequate for a market economy. The industry is privatized to a great extent, but its development has been limited by state policy (Table 4).

The new economic realities have led to vertical integration in the agricultural and food chain. Processing plants, especially in the meat, dairy, sugar, and vegetable-processing industries, have begun to contract with agricultural producers on a long-term basis, to invest in farms, and to provide farms with seasonal monetary and commodity loans. The biggest processors tend to buy small regional processing plants in order to use them as initial raw material collectors or initial processors.

Table 4. Share of Privatized Enterprises in the Processing Industry, January 1, 1997

(percent)

Industry	Share of privatized enterprises in the total number	Share of privatized enterprises with partial state ownership	Share of privatized enterprises in which farms have a controlling interest
Total processing industry	92	18	14
Meat	92	16	9
Dairy	92	18	16
Elevators and mills	90	35	14
Flax processing plants	83	32	11

Source: Ministry of Agriculture and Food survey in 72 territories of the Russian Federation.

Experience in Three Regions

The processes in the downstream sector described above have significantly changed the food chain. It has become more diversified, and new players have appeared, although barter deals occupy too large a share in the sector. Table 5 is based on average data from a survey of three *oblasts* (regions) in European Russia. Of these, Pskov had a very poorly developed food chain and greatly deteriorated agriculture; in Orel there was marked state intervention in agricultural and food markets; and Rostov had a rather liberal regional policy and a developed market.

Rostov producers are much more oriented toward commercial marketing channels: there is a more advanced intermediate structure, producers more frequently utilize commodity exchanges and wholesale markets, and there is greater flexibility in the choice of channels. It should be emphasized that individual farms made deliveries to more than one processor, and not necessarily in the same district as the farm. For the Soviet economy, Rostov's example was remarkable.

Table 5. Marketing Channels for Agricultural Products by Share of Gross Output, 1994–95

(percent)

Channel	Wheat	Barley	Potatoes	Milk	Cattle
Processors	39	5	0	75	63
Procurement agencies; consumer coops	5	24	8	6	3
Retailing network	0	1	0	1	3
Town markets; retailing from trucks	0	0	0	8	2
Intermediaries	3	13	12	0	2
Direct deliveries	4	3	6	0	2
Fairs, wholesale markets, exchanges	0	8	0	0	0
Barter	16	5	2	0	4
Payments in kind to personnel	4	2	1	1	1
Sales to households	9	9	31	2	15
Sales to other farms	3	8	0	0	1
On-farm consumption	14	22	40	6	3
Other	1	0	0	1	2

Note: The data are from a survey of 89 large-scale farms in Orel, Pskov, and Rostov oblasts carried out by the Institute for the Economy in Transition and the University of Kiel and led by J. von Braun and Eu. Serova; see von Braun and others (1998).

An analogous survey in the same three regions but for a smaller number of agents and products was undertaken in 1997–98. Intermediaries' share of grain sales had increased significantly, to about one third of total sales. Large-scale farms increased their deliveries of cattle and milk to conventional processing plants in comparison with 1995. Individual producers sold up to three quarters of their beef to intermediaries and about three fifths of their milk to final individual consumers. These changes were mostly brought about by producers' desire to get cash for their products and so avoid depositing the revenues in bank accounts. (Because of the accumulated debts in agriculture, bank accounts for most farms are blocked. In accordance with financial regulations, all revenues flowing into such accounts must be transferred to fuel suppliers and other creditors to cover debts, without the permission of the account holders.)

The survey showed that in 1997 only 9 percent of wheat deals passed through bank accounts; for milk, the share was 18 percent, and for cattle, 14 percent. The preliminary results of the survey in the same three regions in 1997–98 showed that cash and barter deals accounted for up to 100 percent of transactions in cash crops. This trend can be examined in two ways: the cash and barter deals signal market failure, but at the same time they demonstrate the farms' ability to adjust and survive in the context of imperfect markets for inputs, output, and capital.

Credit

Soviet collective and state farms operated within soft budget constraints; loans were allotted to the farms at extremely low rates and were periodically written off. Financial discipline was low. Managers tended to obtain large loans, anticipating that they would not have to be paid back. The major Soviet agricultural bank (Agroprombank) carried out cashier functions; its staff was not equipped to operate in a market environment.

In the initial period of the reforms the only source of seasonal finance for farms was soft centralized credit that was allocated by the privatized but not radically restructured Agroprombank at easy interest rates. The soft credit terms led to a secondary market for loans: farms resold the centralized loans at the market interest rate. Later, the severe budget crisis led to the establishment of a centralized commodity credit system in 1995–96.

In the fall of 1994, for the first time since the start of the reforms, the accumulated debts of the farms were written off. (Actually, they were restructured, but no payments in accordance with the restructuring scheme were made, and the debts can be regarded as expunged.) The amounts were considerable—more than 20 billion rubles—and the writing-off was not linked to any conditions for farm restructuring. This step led to a worsening of the financial discipline of farms, which were just beginning to become aware of real budget constraints.

In the fall of 1996 the nearly bankrupt Agroprombank was sold on tender to SBS, a commercial bank established in the reform period, and SBS-Agro was set up. At the same time, a special fund for soft credit to the agroindustrial sector was created. Two banks, SBS-Agro and Alfabank, were authorized to distribute fund monies to agroindustrial borrowers.[7] In spite of many distortions and interference by state structures, it was the beginning of real banking in the sector.

A survey of the lending practices of SBS-Agro in five typical agricultural regions of Russia shows that repayment of loans is not related to the legal form of the farms. Because of the farms' deep indebtedness, banks allocated the loans not to the farms directly but to the upstream

---

7. For further details about the evolution of the agricultural credit system in Russia during the reforms see the chapter by Serova and Yanbykh in Gaidar and others (1998).

and downstream sectors and to regional administrations. Two years' experience, however, has demonstrated that the borrowers with the worst repayment records are administrations and input suppliers; the best are the processors and traders. A notable development is that agricultural producers have started to repay their loans. Certainly, loans tend to go to farms that have relatively high credibility in the first place, but the changes are also a sign of a new situation in Russia's agricultural credit system.

Banks tend to increase loans to the types of borrower that repaid in the previous season (Table 6). In 1998 administrations and suppliers in the sample regions received much smaller loans than in the previous year, while farms (both large scale and individual) and processors received much larger amounts (Figure 1).

Thus, agricultural credit has developed as a market tool targeted at the most efficient farms, leaving less efficient farms to rely on subsidies from local administrations and to practice survival behavior. The recent crisis in Russia, however, greatly undermines this emerging agricultural credit system, and centralized credit is likely to again become the main source of agricultural finance.

Table 6. Delinquent Debts and Rate of Growth of Loans by SBS-Agro in Five Regions, by Type of Borrower

Borrower	1997		1998		Loans in 1998 as a percentage of loans in 1997
	Loan (thousands of rubles)	Delinquent debt as share of loans (percent)	Loans (thousands of rubles)	Delinquent debt as share of loans (percent)	
Large-scale farms	158,759.3	9	102,718.5	0.5	688.9
Individual farms	6,411.0	6	16,508.4	0.2	4,644.3
Processors	126,416.2	4	113,993.3	0.2	2,403.4
Procurement agencies	81,334.7	1	25,712.4	1.1	6,014.4
Input suppliers	46,550.0	89	4,317.1	0.2	10.5
Other (mainly regional administrations and food corporations)	251,290.4	49	11,085.0	0.9	10.4
Total	670,761.6	28	274,334.7	0.4	162.9

Note: The five regions are Belgorod, Kursk, Lipetsk, Omsk, and Tambov.
Source: SBS-Agro data.

Figure 1. Distribution of Loans from the Soft Credit Fund by SBS-Agro to the Agroindustrial Sector in Five Regions, by Type of Borrower

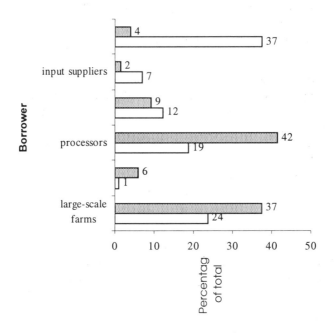

Note: The five regions are Belgorod, Kursk, Lipetsk, Omsk, and Tambov.
Source: SBS-Agro data.

Conclusions

The emergence of market-oriented producers was the primary objective of the reforms in Russian agriculture. Such producers have to operate in accordance with price signals derived from the interaction between demand and supply in more or less freely operating markets. The Soviet collective and state farms had demonstrated their complete inability to respond to market signals and therefore to operate in a market environment.

The reforms in agriculture undertaken in the early 1990s were inconsistent and incomplete. Markets in the sector—for commodities, labor, land, and capital—are rather underdeveloped and nontransparent. The macroeconomic situation stipulates real farm transformation.

Nonetheless, the general economic reforms, together with privatization and farm restructuring in agriculture, have generated a certain amount of progress in the economic behavior of agricultural agents. The sector has become responsive to price signals. Farms are now more independent in their choice of marketing channels, and these channels have become more diversified and more relevant to the market system. Given a sound financial setting, farms demonstrate a financial discipline that is remarkable for Soviet-type producers.

The efficiency of Russia's agriculture remains extremely low. Restructuring of agriculture and the related infrastructure lags behind that in the most advanced transition economies, and the prevailing macroeconomic instability does not favor growth in the sector. Still, some progress toward the creation of a sector of market-oriented agricultural producers has been made. In a sound economic environment, the largest segment of agricultural producers (which determines the performance of the sector) can rather easily revive and operate on a true market basis.

References

Gaidar, Yegor, and others, eds. 1998. *Economics of Transitional Period* [in Russian]. Moscow: Institute of the Economy in Transition.

Ickes, B., and R. Ryterman. 1994. "From Enterprise to Firm: Notes for the Theory of Enterprise in Transition." In R. Campbell, ed., *The Postcommunist Economic Transformation: Essays in Honor of Gregory Grossman.* Boulder, Colo.: Westview Press.

Korbut L., and I. Khramova. 1996. "Main Trends in Food Chain Formation in Russia" [in Russian]. *Business in Russia,* no. 1: 12–21.

Serova, Eu. 1996. The Specific Features of State Support of Agriculture in Russia" [in Russian]. *Issues of Economics,* no. 7: 88–100.

Serova, Eu., and R.Yanbykh. 1998. "Agricultural Credit in Russia" [in Russian]. *Issues of Economics,* no. 11.

von Braun, Joachim, Eu. Serova, K. Frohberg, and P. Wehrheim, eds. Forthcoming. 1998. "Papers presented at the conference 'Russia's Food Economy: Towards Truly Functioning Markets,' July 13–14, 1998."

Survival Strategies of Enterprises and Families in the Contemporary Russian Countryside

Zemfira I. Kalugina

In the summer of 1998 the author directed and participated in a study on the adaptive strategies of agricultural enterprises and families in unstable environments. The study was conducted by the Department of Sociology, Institute of Economics and Industrial Engineering, Siberian Branch of the Russian Academy of Sciences, with financial support from the Russian Humanitarian Scientific Foundation. Four agricultural enterprises were chosen, with expert advice, in two rural districts of Novosibirsk *oblast*.

The sources of information were interviews (both focused and unstructured) with 60 enterprise managers and specialists; a questionnaire survey of 404 employees; an analysis of data for 1992–98 (the period of radical economic reforms); and documents and literature in the field. The objectives of the project were to carry out a comparative analysis of the adaptation strategies used by economically strong and economically weak agricultural enterprises, evaluate the effect of these strategies on the economic behavior of workers and their families in primary and secondary sectors of the economy, and assess the economic performance achieved by the strategies and the social costs they involved.

Three aspects of adaptation strategies were analyzed: economic strategy, social policy, and manpower policy. *Economic strategy* included choice of organizational and legal status, diversification of the commodities produced, innovation policy, presence of processing facilities and off-farm opportunities, and policy with regard to subsidiary (household) and private farmers. *Social policy* was assessed according to two criteria: development of social infrastructure and maintenance of employees' living standards. *Manpower policy* was evaluated by looking at changes in the number of employees, the enterprise's standing in the local labor market, youth policy, and opportunities for skill upgrading and education

This approach had a number of advantages for researchers. It made it possible to:

1. Identify special features in enterprises' economic, social, and manpower policy and assess the objective and subjective factors in economic growth and social stability

2. Analyze the conditions and opportunities under different adaptation strategies that affect employees' ability to realize their labor potential

3. Assess the social costs of the chosen strategies for the enterprise and for its employees and their families

4. Evaluate the degree of social tensions among the personnel, as well as employees' involvement or alienation

5. Identify the means used by employees and their families to maintain and improve their living conditions, offset social costs, and provide for the family's future under the different adaptation strategies pursued by agricultural enterprises

6. Review the outcomes.

This paper focuses primarily on the last two point.

Economic Strategies and Forms of Adaptation of Enterprises and Families

A distinctive feature of the present period of adaptation for Russian enterprises and households is that—concurrently with the appearance of new institutions, new processes, and new social phenomena—existing forms are being restructured. Adaptation as "response to innovations" is taking place alongside adaptation as "response to transformation" (Underwood 1954: 372). Korel (1997) notes that adaptation can also be a response to a change (a novelty or alteration) occurring in the internal structure of the adaptant (the adapting person). After the sweeping reorganization of agricultural enterprises enforced in Russia in 1992–94, it is important to know how economic agents have adapted to their new status, functions, and roles. In this study the adaptants are agricultural enterprises and rural households, and the economic strategies of enterprises can be viewed as a means of coping with a radically changing social and economic environment.

Four types (models) of economic strategies used by agricultural enterprises and households in the unstable Russian environment have been identified. (See Annex Figures A-1 and A-2 for households' goals and their assessments of the changes.)

Model 1. Active market strategy: productive, innovative adaptation of the enterprise; orientation of most families toward improving or maintaining their present living standards

Economic strategy and outcomes. This economic strategy is marked by an active, innovative policy; introduction of advanced technologies; cooperation with scientific institutions and with foreign firms; development of processing and storage facilities; effective work incentive schemes; high-level organization of primary and auxiliary activities (in particular, repair of farm equipment); and stable links with business partners that may include arrangements for marketing the commodities produced. This type of adaptation is characterized by clearly defined objectives and by use of new, efficient methods and measures (Korel 1997).

Although the unfavorable economic conditions throughout Russia's agrarian sector affect all economic agents in nearly the same way, in each Russian region it is possible to find agricultural enterprises that are functioning successfully in spite of the present difficulties. What is remarkable about such enterprises is that they are adapting quickly to new economic conditions: they are exploring the market situation, restructuring their production according to market demand, finding the best channels for selling their products, and setting up their own networks of shops. Some of them are founding large commercial firms and creating modern agrobusiness companies. But according to estimates, such innovative enterprises make up no more than 5 to 7 percent of agricultural enterprises in Russia (Poshkus 1997: 14).

In the particular enterprise surveyed, efficiency of production is 40 percent, and receivables exceed payables. Over the period of market transformation (1992–98), the area of farmland was reduced only slightly. The absolute livestock head count decreased by 20 percent (cows by 10 percent); productivity decreased somewhat but still surpasses the average level in the region.

Social policy. Enterprise social policy is characterized by maintenance and development of rural social infrastructure. The enterprise supports its shareholders' and employees' subsidiary farming, organizes summer vacation activities for children, and carries out landscape improvements in the community.

Manpower policy. Not only is labor retained; during the period 1992–98, it increased by 11 percent. There is a contract system of labor relations, regular cash payments of wages, incentives (50 percent extra payment) for high performance and economical use of resources, and punitive dismissal from the company for breaches of discipline.

The last point is especially significant: losing one's job means not only loss of wages but also loss of access to the resources necessary for subsidiary farming. Workers receive these resources (grain, hay, young animals, and the like) either as in-kind payment of wages or at reduced prices. In addition, machinery operators have at their disposal agricultural equipment and transport vehicles, which they use to render paid services and derive income. "The greater the difference between the total sum of goods and services obtained directly or indirectly by the job holder from his enterprise and his 'nominal' wage on the payroll sheet, the more 'valuable' is this job" (Fadeyeva 1999).

The company allocates funds for young people's schooling in institutions of higher education, paying 7,000 rubles per student per semester, and organizes business travel abroad for its employees. Men coming from military service are given the cash equivalent of 5 times a minimum monthly income on their settlement. The chairman of the company advises that young people buy shares in the enterprise. Retired workers receive 3,000 to 7,000 rubles, which amounts to about one or two annual incomes.

Household strategies. Households seek to improve the current level of living standards (26 percent of families) or to maintain it (49 percent), primarily through small-scale commercial farming and secondary paid jobs. Seventy-six percent of the households operate small subsidiary farms; 35.5 percent of the respondents would like to have a secondary job. Over 60 percent prefer to work overtime on their primary job to increase their incomes. Regular payments, if only of a modest cash wage plus income from a small farm, can provide an acceptable living. Over a fourth of the respondents in this enterprise are quite satisfied with their living standards (see Annex Figure A-3).

Opinions expressed by company managers and specialists in interviews

"Our company 'prospers' only in comparison with other companies."

"If the money leaving the country for 'Bush legs' [fried chicken legs] were given to our agriculture, it could really prosper."

"Dividends—we don't talk about dividends! We would be satisfied at least to keep up production, pay minimum wages, and pay taxes on time.

"When I was in France, I paid 400 francs for shoes, while meat there cost 80–100 francs for 1 kilogram. So I had shoes for 4 kilograms of meat, when at home I would have to pay the price of a whole suckling pig for these shoes. You see the disparity; what use is it to talk about prospects for the countryside?"

Model 2. Conformity: compensatory adaptation of the enterprise; dominant household strategy aimed at maintaining the current standard of living

Economic strategy and outcomes. Under this model, the firm uses a compensatory strategy of adapting to the environment by shifting to another mode of functioning (Tsaregorodsky 1975: 177). For example, the surveyed firm, a former state farm that produced meat, dairy products, and grain, was reorganized as a closed-type joint-stock company with the same specialties but with ventures into partnership- or cooperative-based agricultural processing. In the present environment, this combination of primary and supplemental activities allows the enterprise to tackle the problems of marketing its commodities, obtaining cash, and realizing higher profits. This method of adaptation ensures steady growth. Such enterprises, even if they do not exactly flourish, are able to sustain operations in difficult economic conditions. We estimate that about 10–15 percent of enterprises in the Russian agrarian sector use this strategy.

The surveyed company has a variety of processing activities. It has built a milk-cooling and -pasteurizing facility on a partnership basis. Plans are to diversify milk processing by producing sour cream, yogurt, and cheese. The mill built by the grain-collecting station is used on a shared basis. Grain is stored for a charge at the grain-collecting station but its marketing and distribution are entirely managed by the company. The company has regular partners in marketing and processing but believes it could increase its profits by having its own retail trade outlet, and one is being built in the district seat. A construction shop does internal jobs and supplies paid services to residents. The farm has opened its own bakery.

Profitability is 11 percent, and receivables exceed payables fivefold. Recurring partners' arrears of payments for supplied products greatly undermine the company's opportunities to develop production without loans. The agricultural land area was increased by leasing 300 hectares. The cattle herd declined by 30 percent over the period 1992–98 (cows by 35 percent), but the number of pigs increased. Livestock productivity has improved. Milk yields were 3,490 kilograms per cow in 1997 and rose to 4,000 kilograms in 1998. The number of tractors decreased by seven, and no new equipment is available. The company deals with this problem by purchasing second-hand equipment or by cannibalizing old equipment for spare parts. Purchases of old equipment are primarily from privately operating farmers who could not survive and closed down their businesses.

In general, the enterprise has managed to maintain a balanced approach. It plans to expand production, improve grain yields, and increase cattle productivity; according to the specialists, there are good opportunities for growth.

Social policy. The company has built a health-care rest home and runs social facilities such as shops, a school, communications, and medical services. The social services workers have transferred their land shares to the company by contract. Working employee-shareholders and pensioners are assisted in their subsidiary farming with hay, grain, other fodder, and young animals in return for their work. (Animal breeders are given calves.) Once every quarter, each employee-shareholder is given 15–20 metric centners (approximately 150–200 kilograms) of grain free of charge; other employees can buy grain at discount prices. Shareholders receive 20–30 centners of hay and unlimited quantities of straw for their privately owned livestock. During the harvest, equipment operators are provided with food at discount prices. The neediest can obtain food from shops on credit against wages. The most acute problem is housing construction.

Manpower policy. No mass dismissals have taken place, and the number of company employees (primarily involuntary migrants from former Soviet republics) has even increased slightly. Young shareholders who settle down are given a one-time cash benefit and are assisted in starting their own subsidiary farms. School graduates are given jobs.

Opinions expressed by company managers and specialists in interviews
"Our innovations? First of all, general efforts, ranging from the director to rank-and-file workers, to economize on material inputs."
"We are fighting small thefts that are still practiced, but their scale is quite moderate."
"Requirements for the workers are very strict; discipline has improved."
"At our meetings we keep persuading people that they are owners, shareholders. But it is not easy to awaken this feeling."
"With support from the company, a man can gain maximum profit from his own household farm, too."
"Before the reform, milk yield was 3,000 liters per cow; the state farm was a millionaire. Many workers could purchase their own cars. We had prosperity then. And they could provide education for their children . . . Now we work better but live 10 times worse."
"At present, the work is very exciting, but very difficult too. Now we have self-reliance: your skills determine how much you gain. It is wonderful!"

Household strategies. The prevailing orientation is to maintain the current standard of living. One out of two workers would like to have a secondary paid job or to work overtime for extra income. About half of the families have additional cash from sale of products derived from the household farm.

Model 3. Mimicry: deprivation adaptation of the enterprise; orientation of families toward survival or maintenance of current living standards

Economic strategy and outcomes. The mimicry strategy is characterized by low innovative activity. Its typical features are limited diversification of production, little development of processing facilities, and lack of off-farm businesses. According to specialists in the surveyed enterprise, an opportunity has already been missed. In their view, the establishment of processing facilities and production should have started six or seven years ago. Expansion of these activities is not planned because of lack of funds and difficulties in obtaining loans. Commercial loans obtained through food corporations are not being repaid because the prices for agricultural products are very low, while the prices of purchased inputs are very high.

The set of protective measures and instruments reflected in the mimicry strategy is directed more toward survival than toward growth. This strategy corresponds to the deprivation model of adaptation, which is based, as a rule, on self-reliance, lower expectations, and low consumption standards (Korel 1997). In our study an enterprise that had been one of the best in the district and oblast declined in all basic dimensions. After the reorganization and the attempts at market transformations, this situation has become characteristic of more than half of all agricultural enterprises.

The enterprise in the case study has had a dairy under construction for four years and has invested large amounts of resources in it but is unable to complete the work because of lack of funds. A bakery has been opened, and there is a small carpentry shop, but with only two workers.

The livestock herd, including cows, decreased by about a third between 1992 and 1998, and livestock productivity dropped by about a fourth. The cultivated area has not changed. Profitability over the period under consideration fell to a third of the original level and in 1998 was 10.6 percent. Payables exceed receivables 15-fold. In 1997 the enterprise showed a loss of 3,000 rubles.

According to the specialists, 80 to 100 percent of the vehicles and other agricultural equipment is worn out. Since 1992 the number of tractors has decreased by five, and the number of combines by three. The last purchases of new equipment took place five to seven years ago. Second-hand equipment has been purchased, partly on credit.

Social policy. The company has preserved all of its social infrastructure—the music school, kindergarten, and House of Culture—but housing construction has been practically frozen. The situation in small villages is much worse: in some of them the school, clubhouse, and shop have closed.

In case of urgent need, workers are given cash loans at 10 percent interest. The loan of about 1,000–3,000 rubles usually has a term of a year or a year and a half. As a rule the money is spent on current consumption rather than on purchase of durable goods. If wages are delayed (they are often two months in arrears), workers draw goods from the shop against wages.

Manpower policy. The average number of workers on the payroll was reduced by 27 persons between 1992 and 1998. According to the specialists, one reason for manpower turnover is the expanding urban-rural gap and the low incomes of agricultural workers compared with other sectors of the national economy. The average monthly pay in 1997 was 498 rubles, which is below the subsistence minimum. An absolute majority of rural residents have to maintain small household farms, thus taking on a heavy physical workload. The young people seek either to leave the countryside or to find a nonagricultural job on the local labor market. The company could attract them by providing new housing, but that is at present beyond its power.

Although workers' skills are generally satisfactory, there is a need for constant skill upgrading. If people do not travel out of their areas and do not meet other people, it is difficult for them to maintain their skills at the standard professional level. Specialists are losing their skills, as most courses now have to be paid for and the company cannot afford travel for its employees.

Household strategy. Household strategy is aimed at survival (40 percent of families) or maintenance of current living standards (42 percent); 16 percent of families seek to raise their living standard through small household farming and secondary paid jobs (Annexes Figure A-1).

> **Opinions expressed by company managers and specialists in interviews**
>
> "No liberalization has occurred in the countryside. Everything has remained as of old: one administrator, several specialists. All power is with the enterprise director; employees are under the director's authority and guidance. While nearby state farms are ruined, ours is still functioning."
>
> "The director makes every effort not to let the enterprise be ruined and to let people live in peace, be employed and socially protected, and maintain their high earnings and material welfare."

Model 4. Passive biding: destructive adaptation (disadaptation) of the enterprise; family survival strategy

Economic strategy and outcomes. Typical features of this economic strategy are reorganization of the enterprise into a closed-type joint-stock company; passive biding in hopes of changes and of aid from the top; total depreciation of fixed capital; absence of own processing facilities and of stable ties with customers or mediators; and little diversification of production. Such a "strategy" reveals that the enterprise has been unable to find its niche in the new economic space and to adapt to the new economic environment. The result is disintegration of the staff and eventual ruin. This "destructive adaptation" prevails, in our estimation, in one out of five companies.

The surveyed enterprise had specialized in poultry production. At present, it is a grain, meat, and dairy producer.

Livestock decreased by 75 percent between 1992 and 1998 (70 percent for cows), and cattle productivity declined. In 1998 annual milk yield was 1,628 liters per cow, which is below the average for the region. During the period the number of tractors fell by 104 and the number of combines by 38. Four or five years ago, the equipment was leased out, but it was not paid for, and now it is in disrepair. Payables exceed receivables more than twofold.

Social policy. The company's social policy is aimed mainly at assisting employees and pensioners in their subsidiary farming, rendering services to residents at discount prices, and solving transport problems encountered by the school and the medical facilities. The reduced opportunities available for ameliorating social problems intensifies social tensions among the personnel and causes a great many conflicts.

> **Opinions expressed by company managers and specialists in interviews**
>
> "No help from anywhere; we are no one's concern."
>
> "No ideas, no innovations."
>
> "All has been ruined; equipment is in disrepair."
>
> "Of young people, only odd persons come, undisciplined, drinking, those whom nobody needs."
>
> "Workers are stealing all things of use, pilfer unused buildings, fodder, equipment – everything."
>
> "People live according to the old understanding—that it is not mine but everyone's."
>
> "It is a deadlock."

Manpower policy: Between 1992 and 1998 the staff decreased by more than half, and dismissals continue. For a long time the remaining workers have not been paid regular wages but only small cash advances. In 1998 wages were paid only occasionally. Before the reorganization of the farm, no wage arrears were experienced.

Household strategies. The result of the destructive deprivation model of adaptation is that employees barely manage to subsist. Over 70 percent reported deterioration in their material situation in the past two years (see Annex Figure A-2). The chief source of income, and of income in kind, in particular, is small-scale farming, which greatly increases the employee's workload; 40 percent of respondents said they work to the point of exhaustion.

Domestic strategies are oriented mostly toward survival, even extending to illegal actions. A majority (89 percent) of rural respondents excuses illegal behavior, fully or partly, because they see no other way out. About half of the respondents attribute their disastrous situation to the poor financial standing of the company. Half of the employees could not recall any happy events in their lives in the past year, and another 5 percent said that the past year had brought them only trouble.

What helps these families to survive is the high commercial profitability of their small farms. The share of small household farms for this company is the highest in our survey: 77 percent. Most respondents expressed willingness to have an additional paid job.

Conclusions

Notwithstanding the similarity of external conditions, the transformed agricultural enterprises exhibit drastically different models of strategies for adapting to new social and economic conditions.

The choice and results of a particular adaptation strategy are determined to a great extent by the personality of the enterprise manager. The analysis showed that the most successful enterprises are those that did not have a change of leaders during the reorganization period. They are headed by so-called "red" directors who are not ideologically sympathetic to the present course of reforms but, in practice, follow market models of economic behavior. By contrast, young leaders, while embracing and supporting the ongoing transformations, do not have experience with practical work and find themselves helpless in the face of the difficulties of the present economic environment in Russia.

According to 40 percent of the rural people surveyed, the deterioration in their enterprise's economic situation is caused by the leader's lack of skills and erroneous actions. About the same share of rural respondents attributed their troubles to the erroneous course of reforms, ill-prepared state policy toward the agrarian sector, and a faulty tax system. The workers associate hopes for their enterprise with changes in the agrarian, tax, financial, and credit policies of the state and with the replacement of their enterprise manager. One in five employees thought it necessary to raise the work motivation of rural people and improve work discipline. It is revealing that only 2 percent of respondents hoped for a return to the former system and that less than 1 percent thought that their enterprises could not be revived at all. (See Annex Figure A-4 for respondents' assessment of enterprise problems.)

After transformation, the economic situation of all agricultural enterprises deteriorated. Nevertheless, enterprises specializing in production that is profitable under current conditions (for example, grain production) are in a better position than those concentrating on livestock. Animal production of all kinds is unprofitable, given the present price disparities. The economic background of the enterprise—its starting level at the beginning of the reforms—influenced the process of reorganization of the enterprise and its adaptation to new conditions but was not always the decisive factor in its financial situation.

The dominant factors in enterprise survival in the complicated environment are diversification of production, a moderately innovative policy, and development of small processing facilities. Agricultural enterprises seek to encourage local processors for several reasons: to resist the monopoly pressure of large processing enterprises that dictate conditions but do not always fulfill obligations; to ensure a dependable, even though small, source of cash; and to improve the profitability of processing activities by eliminating the middleman.

Most enterprises have no reliable and well-run channels for marketing commodities that offer stable means of replenishing finances. In the food market, conditions are often dictated by second-hand dealers, racketeers, and criminals, and, consequently, barter trade is common. In Novosibirsk oblast the share of barter in the total volume of grain transactions was 22 percent in 1998; in 1992 it was only 2 percent.

Depreciation of agricultural equipment and lack of finances for purchase of chemical fertilizers and for preventive veterinary services lead to failure to observe technical requirements and keep up the operating standards needed to maintain product quality.

The main reasons for the unprofitability of agriculture are the price disparity between agricultural and industrial products and the high cost of energy. These circumstances hamper efforts to solve the problem of agricultural growth and lead to contraction of social programs and housing construction. All this increases social tensions and conflicts in collectives. Still, even in the present unfavorable conditions, enterprises manage to perform many social functions that help maintain their workers' living standards and to assist them with their small-scale farming.

There are no relevant provisions for regularly upgrading workers' and specialists' skills. Charges for education and the higher cost of living in urban areas make it difficult for rural young people to get secondary or higher professional education in the city. The situation is made worse by the absence of opportunities for exchange of experience and contacts among specialists.

The adaptation strategies used by agricultural enterprises greatly affect the dominant household strategies. The main survival methods for rural families under present conditions are to pursue small household farming, take secondary paid jobs, and engage in illegal practices. In economically strong enterprises most families are oriented toward material affluence or maintenance of current living standards. In economically weak enterprises households focus on subsistence.

In many regions of Russia the problems of bankrupt companies are expected to be resolved through improvement programs designed to preserve the production potential—the land, physical, and manpower resources—of insolvent companies. Such programs would create, within the framework of present legislation, new agricultural organizations that will be able, with organizational and financial support from the local administration, to develop in the present environment. Experience in the Novosibirsk oblast shows the advisability and efficacy of such efforts.

ANNEX

Survey Results

Table1: Enterprises and families' intentions

	Model 1	Model 2	Model 3	Model 4
To live better	2	5	2	3
To maintain the present living level	26	15	16	6
Simply to survive	49	44	42	25
Uncertain	23	36	40	66

Table 2: Assessments of changes in their material status

	Model 1	Model 2	Model 3	Model 4
Material status better	21	6	6	3
No change	41	43	32	22
Material status worse	36	47	62	72
Uncertain	2	4	0	3

Table 3: Social Situation in Work Teams

	Model 1	Model 2	Model 3	Model 4
Satisfaction with job	62	47.5	44	38
Conflicts with administration	30.1	35	52.4	53.8
Satisfaction with job	26	7.5	9.4	6.2

Table 4: Major Problems with Work

	Model 1	Model 2	Model 3	Model 4
Non payment of wages	1.1	20	61.9	30.8
Threat of unemployment	1.1	2.5	7.1	9.1
Low earnings	26.9	28.8	29.8	30.8
Obstacles to improved efficiency	39.8	13.8	29.8	16.2

References

Fadeyeva, O. P. 1999. "A Siberian Village: The Alternative Model of Adaptation." In V. Danilov and T. Shanin, eds., *Peasant Science: History, Theory, Present Time. Yearbook* [in Russian]: 227–40. Moscow.

Korel, L. 1997. *Sociology of Adaptations: Essays in Defense* [in Russian]: 40–113. Novosibirsk: Institute of Economics and Industrial Engineering, Siberian Branch of the Russian Academy of Sciences.

MAAMiK. 1997. "Reformation of the Insolvent." In *Reformation of Insolvent Agricultural Organizations* [in Russian]: 3–10. Novosibirsk.

Poshkus, B. 1997. "Inner Reserves of the Russian AIC." *AIC: Economics, Management,* no. 3: 11–19.

Tsaregorodsky, G. I., ed. 1975. "Issues of Philosophy." In *Philosophic Issues of Adaptation Theory* [in Russian]. Moscow.

Underwood, G. 1954. "Categories of Adaptation." *Evolution* 8.

Changes in the Everyday Activities of Rural Women in Russia from the 1970s to the 1990s[*]

Olga V. Artemova

The dramatic economic and political transformations that have taken place in Russia in recent decades cannot but have affected rural life. The rural population, however, responds to change at its own pace and in its own way, its actions in part determined by the seasons, crop cycles, the needs of livestock, and the existing relationships between city and country and between farmers and consumers. This paper looks at how reform and change in Russia have affected daily life in rural areas and, in particular, the lives of the women who carry so much of the workload and have to cope with the effects of social and economic change at the household level.

Everyday activity is a complex of actions directed toward satisfying the most urgent (Maslow's "basic") needs. It is conducted under the conditions prevailing at the family and community levels, it is relatively autonomous, and it depends only indirectly on macroeconomic and political developments. Although this activity is polyfunctional, the number of needs it satisfies other than the basic ones is small. The activity takes place in time, which is its fundamental resource, and can be quantitatively measured. A characteristic feature of women's everyday activities is the complicated structure of the work, with its unpaid, nonmarket component and diverse roles: wife, childbearer, mother, teacher, worker, homemaker, and practical nurse.

Becker (1965) regards the family not only as a consumer but also as a producer. He offers a model that represents the process of optimization of the distribution of labor for satisfying the needs of the family and its members. Toffler (1983) calls "prosuming"—unpaid work aimed not at exchange but at producing goods and services—"a key factor in the new economy," the economy of the postindustrial society. In these statements, time is central because it is a fundamental resource, a quantitative indicator of human activity.

---

[*] The analyses described in this paper are based on research projects 98-06-80149, funded by the Russian Foundation for Basic Research, and 98-03-04014, funded by the Russian Humanitarian Scientific Foundation.

G. S. Strumilin, in his pioneering studies of workers' and peasants' families in the 1920s and 1930s, revealed the enormous opportunities inherent in the time-budget method for studying everyday life. He was the first to employ this method in the study of social change (Strumilin 1923, 1924, 1925). Strumilin believed that in the study of everyday activity, it was necessary to take into account the physical environment and time expenditure, as well as income and money expenditure. He therefore made practical calculations of the amount and results of nonmarket, unpaid household work (see Strumilin 1982 [1932]).

Time input is one of the most important areas of changes in society and provides exceptionally rich and diverse information about the current state of people's life activity. "In the time budget, not only the division of labor in a family, but also the worker's tastes and needs and his general culture, stand out vividly, as they never do in an income-expenditure book" (Strumilin 1982).

The studies most comparable to Strumilin's, in methods, sample populations, organization, instruments, and inclusion of two main seasons, are the time-input surveys conducted in Novosibirsk *oblast* (region) by the Institute of Economics and Industrial Engineering, Siberian Branch of the Russian Academy of Sciences, under the direction of V. A. Artemov.[1] The three time points selected reflect different periods in Russia's development: the growing crisis (1975–76); the first years of the changes (1986–87); and the period of the rapid drop in production and living standards, with significant changes in politics and ideology (1993–94). The fourth summer survey was carried out in June 1999; the fourth winter survey was planned for November 1999. Table 2 tabulates responses regarding self-assessment of family living standards over the period.[2]

---

[1] Novosibirsk oblast is centrally located in western Siberia and is close to the rural regional average on most socioeconomic indicators. The SB RAS team was the first to make use of computers and the cluster analysis algorithm and of the typology of rural communities developed under the direction of T. I. Zaslavskaya (see Zaslavskaya 1980) in the formation of a sampling population of objects (in this case, rural villages) for an empirical sociological study. For a more detailed description of the procedure used to select communities for time-budget studies, see Zaslavskaya and Muchnik (1977) and Artemov (1979).

[2] The data were obtained by the time-diary method. This method is retrospective: respondents are asked the time allocation for the previous day. Experience shows that the time-diary method gives reliable measurements and that other methods are unsatisfactory. (Juster and Stafford 1991).

Table 1. Time Budget of Agricultural Workers

(hours per work week, adjusted for seasonality)

Activity	Men			Women		
Working time	1975–76 (N = 635)	1986–87 (N = 475)	1993–94 (N = 525)	1975–76 (N = 821)	1996–87 (N = 726)	1993–94 (N = 689)
Time related to work (travel)	60.4	60.1	54.7	48.1	45.1	40.1
Household obligations and errands	3.8	4.5	4.3	5.1	6.1	5.5
Work on the household plot	4.0	3.0	3.6	20.8	24.2	24.8
Child care and contacts with children	9.8	15.4	16.3	16.8	18.9	19.8
Playing and talking with children	1.2	1.7	1.5	3.4	2.7	3.6
Personal needs	0.6	1.0	1.4	1.0	0.8	1.3
Sleep	66.3	60.9	62.3	59.1	58.5	58.8
Free time	55.5	48.9	49.0	49.1	46.9	47.1
Education and self-education	22.0	21.9	23.2	13.9	11.7	14.4
Watching television	0.0	0.0	0.0	0.0	0.0	0.0
Entertaining or visiting friends	4.7	9.4	11.7	4.3	5.5	6.2
Reading	4.5	2.8	3.9	2.7	2.2	4.1
Movies, other entertainment	3.1	2.0	2.0	0.8	0.4	0.9
Active rest and sports	1.5	0.5	0.0	1.5	0.1	0.2
Other	2.4	3.2	1.6	0.3	0.8	0.2
Total	0.5	0.5	2.1	0.8	0.8	1.1
Total workload, including child care	168	168	168	168	168	168
Free time, including contacts with children	78.6	83.7	79.0	93.2	96.2	92.5
	22.6	22.9	24.6	14.9	12.5	15.7

Note: The work week is the average of a summer (June) and a winter (November) week.

To obtain a relatively complete picture of the aggregate activity and living conditions of the family, a special study would be needed. In our opinion, such research would be extremely useful, both practically and scientifically. The immediate and delayed consequences of all the changes taking place in a society and state can be adequately judged only on the basis of how the family lives.

The general results of the longitudinal survey follow. For some measures, comparisons with findings from a similar study of Finnish rural women are available.

Table 2 Self-Assessment of the Family's Material Standing by the Working Population in Rural Areas

(percentage of respondents)

Assessment	November 1986 (N = 596)	June 1987 (N = 604)	June 1993 (N = 618)	November 1994 (N = 594)
We can afford everything; there is enough money for everything	10	10	1	9
Overall, we have enough money	41	39	13	
There is enough money for necessities only	39	41	45	49
There is not enough money for necessities	8	9	41	42
Uncertain	2	1	0	0

Work

Surveys conducted over the past two decades show that rural women's workload was 12–18 hours per week longer than men's and that they had correspondingly less free time for rest, creative activity, and self-education (Tables 4 and 7). In 1986–87 the workload peaked, especially among women employed in agriculture. Apparently, in the summer of 1987 women's total workload reached the physiological limit; by 1993–94 it had decreased somewhat. By comparison, the total workload of farm women in Finland at the end of the 1980s was much less than that of Russian women. The percentages of men's total workload were about the same (113 percent in Finland and 116 percent in Russia). The Finnish women's free time, however, was 88 percent of the men's, whereas for Russian women the figure was 53 percent (Niemi and Paakkonen 1990: 92–95).

The "source" of time for women's increased workload was their free-time activities, sleeping hours, and work time. The work time of rural women decreased in the 1970s and 1980s and especially sharply at the beginning of the 1990s. (Its share in the total workload was 55.5 percent in 1975–76, 52.9 percent in 1986–87, and 46.5 percent in 1993–94.) This trend is clearer among women employed in agriculture.

Table 3. Time Budget of the Working Population in Rural Areas

(hours per work week, adjusted for seasonality)

Activity	Men			Women		
	1975–76 (N = 635)	1986-87 (N = 475)	1993–94 (N = 525)	1975–76 (N = 821)	1996–87 (N = 726)	1993–94 (N = 689)
Working time	54.4	54.5	49.6	43.9	43.4	36.5
Time related to work (travel)	4.1	4.4	4.8	4.2	4.6	4.2
Household obligations and errands	5.0	5.0	5.4	23.3	24.5	25.7
Work on the household plot	9.4	13.9	17.1	12.9	15.2	18.6
Child care and contacts with children	1.5	2.6	1.7	3.7	4.3	3.7
Playing and talking with children	0.8	1.5	1.1	1.2	1.3	1.3
Personal needs	66.8	63.2	63.3	61.5	60.9	61.2
Sleep	55.7	51.5	49.7	51.0	48.9	48.6
Free time	25.9	23.7	24.3	17.7	14.3	16.4
Education and self-education	0.5	0.2	0.3	0.6	0.2	0.1
Watching television	6.2	10.3	11.9	4.6	5.6	7.3
Entertaining or visiting friends	4.8	3.0	3.8	4.3	3.1	3.6
Reading	3.3	2.3	2.3	1.8	1.8	2.1
Passive rest	2.8	1.6	1.6	1.8	1.2	1.8
Movies, other entertainment	1.2	0.7	0.1	1.5	0.4	0.1
Active rest and sports	2.6	3.0	1.6	0.8	0.4	0.3
Other	0.9	0.7	1.8	0.8	0.8	1.7
Total	168	168	168	168	168	168
Total workload, including child care	73.6	78.9	77.5	86.8	90.7	87.4
Free time, including contacts with children	26.7	25.2	25.4	18.9	15.6	17.7

Note:. The work week is the average of a summer (June) and a winter (November) week.

In 1987, 14 percent of the working women surveyed had a secondary paid job; in 1993, the figure was 16 percent, and another 27 percent wanted a secondary job because of a sharply felt need for money. Jobs, however, were not plentiful in the countryside, and in recent years there has been a deficiency of job openings. The share of work in the public sector diminished, partly because of lower job satisfaction.

In the period under consideration, notably in the second part, an intensive process of active redistribution of time from the public sector to the family economy was going on. In the total workload, the share of unpaid work by women (for household obligations and private farming) increased from 44.5 percent in 1975–76 to 53 percent in 1993–94. For men, the increase was from 19.5 to 29.2 percent. The output of household plots became more significant for the family, both relatively and absolutely. The difference in time spent working on household plots by women employed in agriculture and by other working rural women declined considerably (Table 3); the time spent by working men showed the most significant increase, 82 percent. The proportion of men and women who would like to curtail the size of the plot or to abolish it altogether decreased substantially. As Kalugina (1991) observes, in general, private farming on household plots is "a stable, organizationally formalized source of self-provision."

The increased amount of time both women and men spent on household obligations stemmed in part from their desire to have better prepared meals and more domestic comforts in the 1990s. It was also motivated, however, by money problems, the difficulty of obtaining convenience and other foods, and the deterioration of the quality of consumer services, especially repair services.

In general, women, more so than men, viewed household work as important partly as a form of cooperation within the family (29 percent of women and 20 percent of men gave this response). Men and women assign an approximately equal value to the other functions of household work. The significance of household work as a way of saving money has increased sharply. In 1986, 25 percent of working women (23 percent of housewives) and 15 percent of men mentioned this consideration; in 1993 these numbers were three to five times higher. Thus, the common view is that the greatest value of household work at present is that it enables the family to be economical with money, which is scarce in the countryside. Seven percent of rural women and 9 percent of rural men said they derived no satisfaction from household work.

When women are overworked, there can be great negative consequences. These include effects that are "postponed" to future generations and that have to do with women's maternal role and their predominance among educational and medical workers.

The family, on the whole, seeks the optimal relationship between the work activities of its members and its main functions: having children and providing appropriate conditions for their health, upbringing, and education, to preserve the valuable human capital potential of its members. Division of labor among family members, covering the range from outside employment to everyday tasks, allows the family to maximize, increase, or diversify its income and enables members to satisfy, to a greater or lesser degree, other, less tangible interests (education, leisure, sociopolititical activity, health-supporting behaviors, socializing within the family, and so on). At present, the economic factor has much the strongest influence on the choices made by rural Russian families and their individual members regarding employment.

The family, of course, is not merely a production and consumption cell. This function, however, along with the moral support and solidarity the family offers, is very important in times of turmoil such as the current period. In this context the opportunities for personal growth and the high sense of stability and security the family provides are positive elements—although there is a large proportion of women for whom this is not the case.

The current period should be perhaps seen as a continuation of the reform period of the 1980s, when there was a disparity between the quantity and quality of an individual's contribution and the remuneration received. At present, the disparity has increased, in several directions. First, intellectual work and work related to family services began to be compensated at a lower rate, while pay in the management and financial sphere and in monopolistic industries increased sharply. Second, pay increases (even if only nominal) are now less tied to the results of labor (the quantity and quality of goods and services). Third, there are great opportunities to gain income from informal activity rather than from traditional employment.

The studies have shown a distinct trend among women toward reduction of time spent working in the public sphere. This trend is associated with the need for a more balanced distribution of time between women's two most important functions: work outside the home, and human reproduction. The first attempts to exercise a choice among the forms of activity on family provision have been noted, both implicitly and explicitly. Activities in these spheres are not mutually exclusive but, rather, mutually complementary trends marking the more prominent role and significance of everyday life and the satisfaction of family needs. In women's eyes, however, the value of household obligations and of work on the household plot did not increase. The reason is the forced character and heaviness of this work.

"Normalization" of activity had both negative and positive consequences for the transition to market relations. A constant adjustment to the consumer market, rising prices, and food shortages has been taking place. The individual-household sphere of work serves as a field for testing and accumulation of "entrepreneurial" qualities.

Free Time and Leisure

Free time is a necessary time space for satisfying various social, physical, and spiritual needs. Leisure, as part of free time, has a reproduction-relaxation function. Over half of the women surveyed believe that the main function of leisure is relaxation; they are a bit less pragmatic than men in their evaluation of leisure functions.

Low incomes in the public sector and nonsatisfaction of the most basic needs prompt additional work in the household and on the household plot. As a result the choice between free time and work on family self-provision favored the latter.

The amount of free time available to the working rural population reached a low in 1986–87 but had increased again by 1993–94, particularly for women. It should be noted that among the rural population, women working in agriculture have the least free time and the highest total workload. The difference in women's and men's free time is significant, about 8–9 hours per week. Over the observation period, it changed very little. Free-time activities mostly have to do with family and information. Watching television became the prevalent behavior: in 1975–76 it accounted for 31 percent of women's free time (for men, 21 percent); in 1986–87, for 47 percent (43 percent); and in 1993–94, for 43 percent (50 percent). The share of the next most popular activity, socializing with friends and relatives, increased markedly, especially among women employed in agriculture. This change was related to the decline in "external" associations because of reduction of travel and formal employment. Women's satisfaction with their use of free time decreased in 1993; one quarter of respondents was content, but one third was not.

The time women spent playing with children did not change, although this activity declined somewhat (compared with 1976) among agriculturally employed women in the summer of 1987, when their workload reached its maximum. In a sense, the prevalent trends toward work redistribution in the family are seen here, too: from increased equality between men and women in "pedagogical" work toward reinstatement of women's dominance in this area.

The material and domestic orientation of people's activities were reflected in how vacation time was spent. In the countryside the proportion of rural women who did not take vacation increased from 18 percent in 1986 to 28 percent in 1994. The share of women going somewhere on vacation decreased from 23 percent in 1986 to 9 percent in 1994. The two main reasons for abstaining from holiday trips were financial constraints and the need to work in the household. Between 1986 and 1994 the share of women who worked during their leaves rose, as shown in Table 1.

Table 4. Activities During free Time

Activities during leave	1986	1994
Worked on household plot	52	60
Procured vegetables, berries, and mushrooms for winter consumption	22	29
Did repair and construction work	34	46
Rested	45	35
Spent time with children	21	18
Underwent treatment	7	4

Note: Respondents may have engaged in more than one of the activities listed.

Respondent's replies concerning how they would like to spend their free time vividly point up urgent needs that were not fully satisfied: relaxation, being with the family, and hobbies—that is, the real "basic" leisure activities.

Table 5. Expected Use of Supplementary Free Time
(percentage of respondents)

Activity	Men			Women		
	1976	1987	1993	1976	1987	1993
Passive leisure, rest	12	10	23	20	11	21
Outdoor activities, walking, fishing, hunting	23	12	33	6	1	11
Hobbies	5	9	7	17	27	27
Reading fiction, novels	21	5	4	32	10	25
Household work	12	15	18	7	10	10
Work on the household plot		6	15		2	5
Interaction with children and family	7	9	7	11	13	18
Movies, other entertainment	9	4	2	18	7	12
Personal hygiene, medical treatment	0	2	1	1	3	3
Sports and athletics	3	10	2	2	3	2
Learning, acquiring new skills	7	3	0	8	1	0

The main trends in time use by the rural population of Siberia in the 1970s–1990s proved to be statistically significant; the nonparametric Mann-Whitney, Kolmogorov-Smirnov criteria were used. Data collected in the surveys conducted by the Central Statistical Committee of the Russian Soviet Federative Socialist Republic help elaborate the picture of the basic trends in the use of time (Table 6). The year 1985 was marked by an intensified motivational crisis. Women reduced the time spent on work and work-associated activities by 3 to 4 hours, compared with 1977. Meanwhile, time spent on household obligations (especially among men) and on child care increased. In 1986–87 the social and psychological situation changed, as people began to hope that hard and diligent work would be reflected in the family's income and material standing (CSC USSR 1978, 1980; CSC RSFSR 1978; CSC 1981, 1986; Goskomstat 1990).

Table 6. Time Budget of Collective Farmers' Families
(hours per week)

Activity	Men				Women			
	1977	1980	1985	1990	1977	1980	1985	1990
Total workload	71.1	69.7	69.9	76.1	89.5	87.3	85.6	87.2
Including working time	49.1	48.4	47.5	48.9	44.6	42.7	41.1	42.1
Time related to work (travel)	5.8	5.8	5.5	4.8	5.1	5.0	4.5	4.3
Household work	6.2	6.3	7.0	11.1	29.4	30.3	32.0	33.7
Work on the household plot	10.0	9.2	9.9	11.3	10.4	9.2	7.9	8.1
Personal needs	68.1	67.9	65.8	66.0	63.1	63.4	63.1	64.2
Free time	26.1	27.9	29.0	22.6	14.0	15.9	17.4	14.1
Other	2.7	2.5	3.3	3.3	1.4	1.5	2.0	1.5
Total	168	168	168	168	168	168	168	168

Note: Data were collected in March. Members of collective farmers' families may have been employed in other branches of the economy.

Source: Central Statistical Committee of the Russian Federation.

Findings

The ultimate objective in a study of people's time input is to obtain valid, reliable, and, if possible, quantitative information about the status of and trends in the activities of individuals, families, and the community at large and in the functions, motives, and preferences relating to those activities. This is especially important in the current period, with its ever more acute need for global changes in activity.

The stability in the total workload of rural women is striking, given the considerable changes in the workload structure as women try to satisfy basic needs with limited resources of physical strength and time. The total workload of farm women according to surveys conducted in different years is shown in Table 2.

Table 7. Gender Workload Comparison

	1923	1934	1973/74	1975/76	1986/87	1993/94
Women's workload (hours per day)	13.2	12.4	12.0	13.3	13.7	13.2
As percentage of men's workload in hours per day	157	112	109	127	121	120

The main change in women's use of time in the 1970s–1990s period was the reduction of work time and the considerably increased time spent on household obligations within a total workload that reached its natural limit in the late 1980s (and that decreased slightly in the first part of the 1990s). The difference in the amount of work by men and by women decreased gradually as a result of the reduction of work time by women and the increased involvement of men in farming household plots.

The diversity of leisure activities has increased constantly, and the communication-information character of these activities has intensified in recent years.

In the everyday activity of rural women, there is a rather large gap between preferences and actual behaviors. The main reason is the forced character of their work.

In our view, three main factors entered into the general changes in the use of time:

- Biological (dangers to physical existence, to present and future health, and to reproduction)

- Socioeconomic (increased inefficiency of the state economy, continued decline in work motivation in the public sector)

- Liberalization of various rights—to a secondary job, to partially paid and unpaid holidays, to allotments of garden plots, to cooperative housing construction, and so on.

There is every reason to use longitudinal studies in certain, mostly typical regions along with national samples (or even in place of national samples if those are lacking). The interpretation of the trends identified as a result of longitudinal monitoring of the conditions and lifestyle of the rural population must rest on theoretical developments regarding such problems as the relationship between the "subjective" and the scientific rationality of everyday activity, and the relationships between everyday life and politics, economics, and ideology under different states of society (evolutionary, crisis, revolutionary, or reformed).

References

The abbreviation CSC refers to the Central Statistical Committee of the U.S.S.R. or of the Russian Soviet Federative Socialist Republic (RSFSR).

Artemov, V. A., ed. 1979. Rabocheye i vnerabocheye vremia selskogo naseleniya *[Work and after-work time of the rural population]. Novosibirsk.*

Becker, G. A. 1965. "Theory of the Allocation of Time." Economic Journal 75 *(299).*

CSC RSFSR. 1978. Byudzhet vremeni rabochikh, sluzhashchikh i kolkhoznikov *[Time budget of workers, employees, and collective farmers]. Moscow.*

CSC USSR. 1978. Byudzhet vremeni rabochikh, sluzhashchikh i kolkhoznikov za mart 1977 *[Time budget of workers, employees, and collective farmers March 1997]. Moscow.*

———. *1980.* Byudzhet vremeni rabochikh, sluzhashchikh i kolkhoznikov *[Time budget of workers, employees, and collective farmers]. Moscow.*

CSC. 1981. Byudzhet vremeni rabochikh, sluzhashchikh i kolkhoznikov *[Time budget of workers, employees, and collective farmers]. Moscow.*

———. *1986.* Byudzhet vremeni rabochikh, sluzhashchikh i kolkhoznikov *[Time budget of workers, employees, and collective farmers]. Moscow.*

Goskomstat. 1990. Byudzhet vremeni rabochikh, sluzhashchikh i kolkhoznikov *[Time budget of workers, employees, and collective farmers]. Moscow.*

Juster, F. T., and F. P. Stafford. 1991. "The Allocation of Time: Empirical Findings, Behavioral Models, and Problems of Measurement." Journal of Economic Literature 29: 471–522.

Kalugina, Z. I. 1991. Lichnoye podsobnoye khozyaistvo v SSSR. Soschial'nye regulatory i rezul'taty razvitiya *[Farming on private plots in the U.S.S.R.: Social regulators and results of development]. Novosibirsk: Nauka.*

Niemi, I., and H. Paakkonen. 1990. Time Use Changes in Finland in the 1980s. *Helsinki: Central Statistical Office.*

Strumilin, S. G. 1923. "Byudzhet vremeni russkogo krest'yanina" [Time budgets of Russian farmers]. Byulleten Gosplana, *Nos. 8–9.*

———.*1924. "Byudzhet vremeni russkogo rabochego i krest'yanina v 1922–1923" [Time budget of Russian workers and farmers in 1922–1923]. Moscow and Leningrad.*

———.*1925. "Byudzhet vremeni rabochikh v 1923–1924 g" [Time budgets of Russian workers in 1923–1924].* Planovoe khozyaistvo, *no. 7.*

———. *1982. Problemy ekonomiki truda [Problems of the economics of work]. Moscow: Nauka. First published in 1982.*

Toffler, A. 1983. *Previews and Premises: An Interview with Alvin Toffler. Montreal: Black Rose Press.*

Zaslavskaya, T. I., ed. *1980.* Metodologija i metodika sistemnogo izuchenija sovetskoj derevni *[Methodology and methods of system research of the Soviet countryside]. Novosibirsk.*

Zaslavskaya, T. I., and V. Muchnik, eds. 1977. "Razvitie selskih poselenij" *[Development of rural settlements]. Moscow.*

The Environs of Russian Cities: A Case Study of Moscow

Grigory Ioffe and Tatyana Nefedova

The Russian word *prigorod* evokes different mental associations from its direct translation, "suburb," as used in the United States. "Suburb" implies the intrusion of the city into the countryside and the ensuing implantation of typically urban amenities in a more rarified residential setting. "Prigorod" suggests the countryside itself, but a countryside affected, often in the direction of gentrification, by proximity to a city. Suburbia typically develops from a central city outward, as large groups of people "eject" themselves from the corporate limits of the city. A Russian prigorod largely develops inward as a rural settlement network contracts and as rural people move closer to a city to take advantage of its opportunities and services (Ioffe and Nefedova 1997: 117–38, 258–65). A barrier effect appears at the city line because of the restrictive policy governing residence permits (the infamous *propiska*) in large cities and because of the absence of a real estate market. Thus, the prigorod is a stepping-stone into a city, not out of it as in the West. Russian cities do expand outward, but typically the city line is simply moved, and Western-style suburbia does not take shape. Only most recently, around the largest cities of Russia, has the situation begun to change somewhat.

Features of the Urban Margin in Russia

These distinctions prompted us to reject the term "suburb" and use the terms "exurbia," "urban margins," and "city environs" to denote the rural-urban fringe in Russia.[1] As discussed below, urban margins in Russia have distinctive features that set them apart from those in the West.

Concentration of Population

Russia has been and is a country of large cities; 53 percent of the population lives in cities and towns with over 50,000 residents. Neither the rural-urban fringe nor the countryside dominates the Russian popula-

---

[1] Fishman (1987) notes that Americans use the word "suburbs" for entities substantially different from those the word originally described. As Jackson (1985) observes, whereas "suburban" "once implied a relationship with the city, the term today is more likely to represent a distinction from the city." Fishman proposed the term "metroland" as more appropriate for the American context.

tion. Of the nonurban population, 23 percent is rural.[2] It is difficult to distinguish suburbanites because the residual 24 percent also includes residents of free-standing small towns.

Frequency of Border Changes

Suburban development in Russia has typically been precluded by fairly frequent expansion of cities' outer limits. (In the West, at least in the post–World War II era, city limits have been much more stable.) Moscow's area was barely 56 square kilometers (km^2) in 1860; by 1984, after several outward shifts of the boundary, the area was 994 km^2. Other large Russian cities also extended their borders many times during the Soviet period, although not as much or as rapidly as did Moscow. Areas newly incorporated into cities have become sites for multistory apartment projects, and in many cases they have become more densely packed with high-rise residential developments than areas located between the downtown and the city line.

Distinct Divisions between Residential Sectors

Residential developments near a city fall into three easily distinguishable categories: satellite towns, rural areas, and recreational settlements where urbanites' second dwellings are located. Satellite towns, which house the bulk of the permanent suburban or exurban population, are dominated by multistory complexes and resemble the residential areas of central cities. The second and third categories overlap, but in both cases an American-style single-family detached house with its attendant amenities is even more of a dream than in America. The average population density around Moscow in the 1980s was 575 persons per km^2, but this figure conceals a striking variation in residential density between satellite towns, where the actual density may be as high as 5,000 per km^2, and the countryside, with barely 50 per km^2. As Jackson (1985: 7) noted, the outer boundaries of Moscow and other typical European cities "abruptly terminate with apartment buildings, and a 20-minute train ride will take one well into the countryside." Such an arrangement is a far cry from the vast suburban residential belts typically encountered in the United States.

Importance of Recreational Second Dwellings

In the West usually only the wealthiest part of the population maintains more than one home. The same is true of Russian *dachas*. A dacha implies a large (up to 0.50 hectare) parcel of land and a more or less so-

---

[2] Logos (1996): 68–70.

phisticated house that, depending on amenities, may be used only in summer or all year round. Most other urbanites who own second dwellings in the countryside have small parcels in collective orchard comradeships or still smaller ones in the form of vegetable gardens. No less than 60 percent of Russian urbanites use some kind of summer dwelling outside a city, with a parcel of land that serves both farming and recreational functions. It is the enormous spread of this phenomenon (rather than the phenomenon itself) that distinguishes Russian exurbia from its Western counterparts.

Demographic Viability against a Backdrop of Rural Decay

Russia has never experienced a stage in which cities decline or stagnate while smaller settlements grow because of net migration gains. The sudden rural population growth of 1992–95 was a short-lived phenomenon driven exclusively by the acute crisis that seized the cities as industry stagnated and state price controls were removed. By 1995 the situation had returned to what had been normal for Russia for over a century: a positive relationship between a settlement's growth rate and its size (although the overall negative rate of natural population increase suppressed growth).

Trends in Russia's recent and even longer-term population redistribution are meaningful only when viewed against the backdrop of the overall spatial pattern of population. Russia is a sparsely settled country, and not only in Siberia, which is "literally torn apart by distance."[3] Many Westerners do not realize how acute the sense of remoteness, the perception of living in the middle of nowhere, can be in areas barely 150–250 km from Moscow and St. Petersburg. The primitive condition of roads and other means of communications creates this perception in the first place, and the sparseness of the network of vibrant urban cores, as well as the long-standing centripetal trends in migration, exacerbate the polarized pattern. In European Russia the average distance between cities with populations over 250,000 is 314 km—twice as much as in Western Europe (158 km).

In Russia densely settled heartlands and thinly settled hinterlands interpenetrate. Like oases in the rural vastness, the urban nuclei have been unable to cast the web of intense social interactions over more than a relatively small part of the interurban space. This stands in contrast to North America, and especially to Western Europe.

---

[3] Sergei Tarkhov, quoted in Lappo and Polyan (1996): 11.

In view of the drastic depopulation of most areas more than 2 hours' traveling time from big cities (in central and northwest Russia and, to some extent, the central black-earth area), the vicinities of the cities appear to be almost the only pockets of viable commercial farming. They are also the only areas where one can find the best features of urban and rural ways of life.

The Role of Agriculture

The importance of agriculture in Russia's urban margins is incomparably greater than almost anywhere in the West except in Canada and Australia, with their uniquely high degree of spatial coincidence of population and fertile land (Bryant and Johnston 1992: 25–26). In Russia such a coincidence is largely nonexistent. What accounts for the heightened significance of exurban (*prigorodnoye*) agriculture is that exurbia has become the only area where human capital is available in sufficient quantity and quality and where rampant mismanagement and alcoholism have not had a large impact on commercial farming. Only the piedmont provinces of the northern Caucasus are the exception to this state of affairs in European Russia.

In contrast to the situation in the West, the effects of urbanization on the output and performance of farms have been overwhelmingly positive in Russia, even as cities regularly swallow large areas of agricultural land. Because of rural decay in outlying areas, the spatial pattern of agricultural land use in Russian provinces resembles Thunian concentric rings, with intensity of land use declining outward from large cities, even though the most fertile land is distant from the center. The evidence is seen in productivity gradients (Ioffe and Nefedova 1997: 228–29) and in specialization, with exurban farms usually emphasizing milk, vegetables, green and succulent feed, and potatoes, while outlying ones focus more on grain.

To sum up, the environs of Russian cities appear to be less visible as population concentrations than their Western, especially American, counterparts; more frequently shifted outward due to repeated changes in city boundaries; and marked by crisper internal divisions of residential space. They host myriad agrorecreational second dwellings belonging to urbanites, and they appear to preserve demographically viable pockets of the countryside against a backdrop of overall sociodemographic decay in rural areas. Finally, they are not perceived as predominantly residential areas, in view of the significance of farmland in the "city's countryside."

A Closer Look at Moscow

In this section we focus on a case study of the environs of Moscow within the confines of Moscow *oblast* (region). The term Moscow urban field (MUF) will be used to refer to the city of Moscow and Moscow oblast together.

Population Density and Growth

In 1996, 6,597,000 people lived in Moscow oblast alone; the Moscow urban field had more than 15 million people, or over 10 percent of Russia's total population. Until 1990 the population of the MUF grew faster than that of Russia as a whole.

Population density in the entire MUF is 325 per km^2, but outside the city it is only 140 per km^2. These densities are similar to their counterparts in such countries as Belgium, Germany, Italy, and the United Kingdom. For Russia, however, such a densely packed space is unique; in no other single territorial unit (republic, territory, or oblast) of the Russian Federation does population density exceed 75 per km^2, while in European Russia it averages 27 per km^2 (approximately the same as in the United States). In the area adjoining the MUF, the lowest population density is in the northwest (Tver oblast), the result of a funnel effect that occurs because the area's population is being "pumped out" jointly by Moscow and St. Petersburg (Ioffe and Nefedova 1997: 229–31).

Although 80 percent of Moscow oblast's population lives in cities and towns, rural population density is much higher than anywhere else in central Russia. Many rural residents are commuters, which heightens the interdependence of urban and rural settlement forms in the region. In fact, two thirds of rural residents have "urban" professions.

For decades migration was the leading component of population growth. About 90 percent of the annual 100,000–120,000 migrants to the MUF settled down in Moscow and in the districts most accessible to it, residential restrictions notwithstanding (MFGO 1989: 43). A step-by-step pattern of migration was characteristic; some of those who desired to enter the city initially settled down in Moscow oblast, found jobs in the city, and become commuters. The administrative restrictions first imposed on migration to the city in 1932 were sidestepped through, for example, marriages (including fictitious marriages), apartment exchanges, and status as a *limitchik*—part of a migrant labor pool whose members were not subject to normal migration restrictions because they undertook nonprestigious jobs scorned by Muscovites. In the 1980s migration into the MUF was much higher than into the adjoining regions.

Spatial Pattern of Development and Population Growth

The principal features of the area's spatial structure are the overwhelming dominance of a single center; a pattern of radial spokes and concentric rings; asymmetric "green west–smoky east" development; and the presence of significant urban-industrial clusters, including those at junctures with other oblasts (*Moskovskiy Stolichnyi:* 72–73).

Grouping all 39 *rayons* (districts) in the oblast according to their proximity to Moscow, we identify four rings, the innermost with an approximately 20-km radius. The differences between the radial sectors are clearly visible in their economic structure, urbanization, and per capita production. According to a typology of districts by Treivish (1996: 22), the city of Moscow is surrounded by service-industrial areas in the north, agroindustrial areas in the south and east, and service-agrarian areas in the west. The southern and eastern sectors accounted for two thirds of the oblast's industrial output in the early 1990s. The asymmetry of industrial development has deep historical roots and has had ecological benefits (*Moskovskiy Stolichnyi:* 78, 265). Airflow is predominantly westerly in the area, and the western sector "supplies" the city of Moscow with fresh air and water and is regarded as the most desirable area for recreation and residence. In the 1970s and 1980s this sector attracted branches of elite Moscow-based enterprises, especially defense-oriented ones. The resulting accelerated diffusion of industrialization caused concern among environmentally conscious people, but the economic crisis interrupted the process. So, the semidepressed east remains the economic buttress of the oblast in a traditional sense, with its concentrations of population, resources, and industrial output, including export-oriented production.

The population declines recorded in the 1990s for both the city and the oblast are associated less with migration than with the reversal in natural population growth. Although this change has been general in Russia, the toll on the MUF's aging population has been heavier than elsewhere. In 1990, for example, natural increase countrywide was 22 per 10,000 population, but it was minus 22 in Moscow oblast. By 1995 the figure for all Russia was minus 57, while for Moscow oblast it was minus 104. Net migration in 1995 could not outweigh the landslide excess of deaths over births.[4]

In the innermost ring, population losses have been minimal, but the rural population has declined almost everywhere. With the onset of the

---

[4] Goskomstat (1996): 711, 733; Goskomstat (1995): 41.

economic crisis at the beginning of the 1990s, population losses in the two most remote rings slowed because many people thought it would be easier to make ends meet in a provincial or semiprovincial environment than near the largest city. However, by 1993 the rates of both urban and rural population decline began to follow a traditional core-periphery gradient. It appears that no Western-style population deconcentration has so far occurred, either when the overall population grew or when it began to decline. The urban way of life continues to be attractive in the oblast. Those unable to enter the central city as permanent residents choose to cast anchor outside Moscow, but usually in an urban apartment complex.

Population flows into Moscow oblast continue to be fairly significant. For example, during the first six months of 1996 alone, net migration amounted to 20,000 people (Filippovich: 24), 8,300 of them from other newly independent states of the former U.S.S.R. Three quarters of all incoming migrants settled down in urban settlements in Moscow oblast; the few who settled in the countryside were not enough to increase population there appreciably.

The bulk of the demand for land is generated by urbanites. Unlike their Western counterparts, they do not seek permanent residency outside city limits but just a second dwelling that can be used for subsidiary farming. The heightened demand for exurban land is generated not only by the sheer concentration of urbanites but also by the traditional attachment to land characteristic of countries of recent urbanization and by the concentration of wealth in the city.

Russian statistics on personal income are unreliable, as people generally report only a small part of their earnings. Even so, statistics show that, on average, a Muscovite is at least thrice as wealthy as the average Russian. While in Moscow 40 percent of residents with gainful (and reported) employment earned over 1 million rubles per month in 1995, only 0.5–6.0 percent did so in surrounding areas, including Moscow oblast. (Note that the official cost estimate for a 19-product consumer basket in Moscow is only 1.5 times higher than in surrounding regions.) The concentration of wealth in the capital city heightens the demand for second dwellings in exurbia and has already created a land and real estate market unique in today's Russia.

The country "estates" of Russian urbanites fall into five major categories: dachas, collective orchards, vegetable gardens, village homes, and *kottedzhi*. They are especially plentiful in the second accessibility ring. Their distribution by sector is more even; only dachas markedly gravitate to the ecologically cleanest west.

Dachas appeared long before the revolution of 1917 as recreational sites for white-collar workers—their version of the estates of the landed gentry. A dacha could be privately owned or rented and was located, as a rule, close to a city, unlike the traditional estate.

In the 1920s government-owned and "departmental" (*viedomstvennye*) dachas sprang up to serve three major segments of the new Soviet elite: the party, economic management, and the intelligentsia. The expansion of private recreational sites, without formal ownership of land, began in the 1930s and continued for about 20 years after World War II. Since then, dachas have largely remained a perquisite of elite social groups. Land parcels for private recreational construction were distributed through workplaces, and *dachniki* were required to be members of a cooperative set up on an institutional basis (e.g., a dacha cooperative of aircraft specialists). The cooperative concluded long-term land-use agreements with local authorities on behalf of its members and provided some maintenance and supply services for dacha owners.

The demand for dachas grew quickly and could not be met. In the 1950s and 1960s the practice of renting parts of dachas and of houses in nearby villages, usually within a 10–50 km radius from the city, became widespread. Initially, dachas were a phenomenon exclusively of the vicinities of Moscow and Leningrad, with ownership of a dacha being a mark of wealth. However, these wooden, two-story recreational dwellings, with 50–80 square meters (m^2) of living space, were gradually simplified, becoming more and more similar to ordinary village homes. In the past the inferiority of rural infrastructure (no plumbing or piped water, and primitive heating devices), even in the vicinity of Moscow, made it impossible to use a dacha year round.

Dacha plots range from 0.12 to 0.5 hectares (ha) and represent the largest-size parcels of land available to urbanites. Initially, agricultural activity on the land was not intensive; forests and meadows accounted for the bulk of land parcels. Many dacha holdings girding Russian cities have retained this picturesque idleness even today.

Prior to the 1990s the land under dachas in the Moscow area belonged to the state, which leased it out through the mediation of dacha cooperatives subordinated to the Moscow oblast soviet. In the 1990s almost all dachas and their land became privatized. Quite a few Muscovites, especially, but not limited to, the elderly, live in their dachas year round while renting out their urban apartments—a kind of hidden suburbanization not reflected in the statistics.

Collective orchards began to be set up in the 1950s, when it became clear that dachas could not meet the demand for second dwellings.

Originally, government branches (such as ministries) and large enterprises leased land from the state and established collective orchards for their employees. In 1980 in Moscow oblast 189 cooperatives or "comradeships" of this kind existed. With the 256 dacha cooperatives, they accounted for 50 percent of recreational land use in the oblast. By 1990, 648,000 families possessed parcels in collective orchards, and by 1996 the number had doubled. The overall land area involved increased from 30,000 ha in 1985 to 110,000 ha in 1996. In most cases inferior land was allotted for collective orchards: disused quarries, marshes, strips immediately underneath high-voltage transmission lines, and other areas inappropriate for collective and state farms. Many parcel owners created topsoil themselves through enormous investments of time and labor.

By 1996, 73 percent of all orchards had been privatized. Most orchards that are privatized are recorded as land belonging to citizens (*zemli grazhdan*), but 25 percent of them are located within industrial production and residential areas.

Initially, dwelling construction on orchard parcels was legally banned, or only primitive shacks could be built. However, the introduction of the two-day weekend in 1967–69 led to a loosening of the restrictions. Tiny houses, unsuitable for living but adequate for spending the night, were constructed. Such parcel-and-dwelling clusters surrounded by ominous fences are the most aesthetically offensive segments of exurbia.

Vegetable gardens are used almost exclusively for agricultural purposes, and most of them lack any built structure. For the most part, they are rented. Individual parcels are smaller than in the other three categories (originally, no larger than 0.01 ha). The number of such parcels has grown rapidly, from 226,000 in 1990 to 557,000 in 1996. Unlike collective orchards, most vegetable gardens are used not by Muscovites but by oblast dwellers. While the center of gravity of collective orchards has been moving outward, most vegetable gardens are located within the first and the second accessibility rings, where urban population density is high.

Only 5 percent of the vegetable gardens are privatized. Two thirds of them are located on industrial and residential lands and one third on the lands of former state and collective farms.

Village homes appeal to city dwellers who are dissatisfied with overcrowded orchard and garden parcels and the associated restrictions. Before 1989, however, an urbanite could not legally obtain a rural house abandoned by its dwellers. The transaction was carried out by making an unofficial agreement with a village *babushka* or her urban relatives or by finding a figurehead rural buyer. In the 1970s a dilapidated but still livable village house could be bought for what amounted to two or three monthly paychecks. By the early 1980s the abandoned houses in villages more or less convenient to a city had been bought up. In 1989 the government issued an official ruling allowing urbanites to buy a house in a village and use the land attached to it. Prices skyrocketed, but the number of those willing to buy a village house continued to increase, and by the early 1990s virtually all vacant village houses within the corporate limits of Moscow oblast had been sold. Muscovites who still wanted to buy village property could do so only outside Moscow oblast, which translated into a three-to-five-hour car trip in each direction.

The number of purchases of village houses peaked in 1990–91. After that the process slowed, but almost every vacant house in Moscow oblast and in parts of adjoining oblasts had been either sold or bequeathed.

"Village" Muscovites abound especially in the west, along the Moscow River and its tributaries, and south of Moscow. In 1993, 16 percent of village homes in Kaluga oblast belonged to Muscovites; in Tver oblast the number was 21 percent. (The overall figure for Moscow oblast is 14 percent, but the share is as high as 50 percent in the most remote districts.)

Urbanites most actively "colonize" depopulated villages; in larger villages with a lot of permanent residents, Muscovites are few. Land and houses bought by urbanites are not recorded as such by land-use statistics, which thus understate the proportion of privatized land. If one takes into account that by 1996 about one fifth of all village houses located in the oblast had already become urbanites' second dwellings (owned or leased) and that therefore about the same proportion of land attached to those houses (up to 0.15 ha) was in private hands, the 421,000 ha of privatized oblast land given in the statistics has to be increased by 20,000 ha.

Kottedzhi (cottages), which appeared only in the 1990s, most closely fit the Western notion of single-family homes. Initially, they were wooden structures on a plot of 0.08–0.1 ha, located close to a city, and they did not exceed $10,000 in price. With more and more "new rich" emerging, the size and price of the kottedzhi began to grow rapidly. Two-to-four-story cottages of stone and brick, looking like stylized medieval castles, were constructed. Demand for these monstrosities, which sold for more than a million dollars, was quickly met and has now subsided. The newest development is townhouse-like two-story cottages grouped in clusters with their own semiautonomous infrastructure and selling for about $100,000. In many cases these houses are suitable for permanent residence. The demand for such cottages, although definitely higher than for "castles," is limited, given that a system of mortgage loans has not yet emerged in Russia.

In 1992, in an attempt to coordinate cottage development, the Moscow city and oblast governments jointly adopted a Program of Low-Story Housing Construction. The original motivation was that about 400,000 people in the oblast and over 500,000 people in Moscow were on waiting lists for improvements in their living conditions (Goskomstat 1996: 159). In effect, the new program was expected to instigate mass suburbanization. Some of its authors envisioned moving up to 3.5 million Muscovites from urban apartments to suburban cottages by 2000 (Zayets 1993: 13–18). About 10 percent of the new dwellings were to be distributed at a gross discount, with the rest to be auctioned. However, the program did not fully take into account the financial situation of many people, and it was soon discredited.

Many bankers and leading industrialists already own cottages, but few ordinary people can purchase such dwellings. In 1995 the offers to sell exceeded demand by a factor of 6 to 20, according to different estimates (Barmina 1995). Luxury cottages are priced at $3,000 per square meter. For a new log cabin more traditional in construction, the price in 1996 was $120–$300 per square meter;[5] that is, even a routine-type structure is not cheap at all.

Agriculture

Milk is regularly delivered to the city of Moscow from as much as 1,000 km away in winter, while vegetables and potatoes come from as far distant as Stavropol in the northern Caucasus.[6] Although Moscow oblast does not play a leading role in provisioning Moscow with perishable foods, it is still one of its important suppliers. In 1995 the city of Moscow required 3,000,000 tons of milk,[7] only 600,000 of which were procured in the oblast; other Russian producers provided about 1,300,000 tons. Of the city's 650,000-ton demand for meat, all of Russia provided only 200,000 tons, about half of which was from Moscow oblast. In 1986–90 the oblast supplied the central city with three times the vegetables, milk, and meat and five times the potatoes it did in 1995.

---

[5] "Skolko seychas stoyat shest sotok v podmoskovye," *Izvestia,* May 9, 1995.
[6] *Mir Novostey,* October 21, 1996.
[7] V. Konovalov, in *Izvestia,* June 2, 1995.

Today, as in the 1980s, yields per unit of land and milk yields per cow are 1.5 to 2.5 times higher in Moscow oblast than in adjoining oblasts, even though areas to the south of the MUF (such as Kaluga and Ryazan) enjoy more favorable natural conditions. In the 1990s the agricultural productivity of Moscow oblast declined more than that in adjoining areas. In part, this was a statistical effect of the low agricultural productivity in the non-black-earth zone: when one cow gives barely 2,000 kilograms of milk (a sheep-like yield), as in the nearby oblasts, there is not much room for further decline.

State and collective farms. To some extent, the productivity slump in Moscow and St. Petersburg oblasts is the result of the high percentage share of public (socialized) farms in their farm output. These former state and collective farms, whose conversion to joint-stock companies has been purely nominal, suffered the most during the crises of the 1990s. Most farmland in the oblast is still in their possession. Moscow oblast contains over 500 socialized farms that specialize mainly in milk, potatoes, and grain fodder; 30 of them are highly industrialized poultry farms. In 1995, 7 percent of the employed population of the oblast worked for such nominally privatized but publicly run production units. This is far short of the share in neighboring oblasts (e.g., 15 percent in Smolensk, 14 percent in Kaluga, and 12 percent in Yaroslavl), but it amounts to about 200,000 people.

The large size of public farms in Moscow oblast is striking when compared with those in adjoining rural districts of neighboring oblasts. For example, according to a 1993 survey, in Shakhovskaya rayon in northwest Moscow oblast, the average public farm had 450 employees, while in neighboring Zubov, Tver oblast, the average number was 120. The national average in the early 1990s was 180 employees.

Subsidiary farms. The role of subsidiary farming by rural residents grew incredibly in the 1990s. Caught between inflation, nonpayment of salaries, and poorly provisioned rural stores, people got by in the old and well-tested way, by working land immediately attached to their homes. In Russia overall, the percentage share of subsidiary farming in total agricultural output increased from 24 percent in 1990 to 44 percent in 1995. In Moscow oblast, with its large and still viable publicly run farms and its relatively accessible urban food stores, the share of subsidiary farms is only half as high as in the surrounding oblasts, but it is still substantial. Subsidiary farming produces one quarter of the total farming output of Moscow oblast, including 77 percent of the potatoes, 28 percent of the vegetables, 34 percent of the meat, and 14 percent of the milk marketed in the oblast. Over the five years 1991–96 the output of subsidiary farms increased by one third.

The total acreage of these tiny plots (0.17 ha on average, comparable to the agrorecreational parcels of urbanites) increased from 60,000 ha in 1985 to 92,000 ha in 1996, mostly at the expense of former state and collective farms. Even so, their total area is about 1/17 that of the state and collective farms. The acreage growth of subsidiary farms in Moscow oblast is far short of that in Russia as a whole (2.5 times), but it is still substantial. In 1996 there were 2.4 rural dwellers per subsidiary farm.

The distribution of subsidiary farmland gravitates toward the second and third accessibility rings. The share of subsidiary farms in total agricultural land is highest in the districts closest to the city of Moscow, while on the oblast's periphery the average landholding is two to three times larger than in the innermost ring.

Private farms

There are about 6,500 private farmers in Moscow oblast, occupying 4 percent of its total farmland. The major difference between private and subsidiary farms in Russia is that the private farms, products of the market reforms of the 1990s, are registered independent businesses (which entails tax- and bank-loan-related consequences); subsidiary farms, by contrast, always existed in Soviet times and are now largely inseparable from state and collective farms, with which they are in symbiotic relationship. The acreage of private farms in Moscow oblast has grown from 500 hectares in 1990, when they first emerged, to 69,500 hectares in 1996. The average private farm's landholding in the oblast (11 ha) is one fourth of the average private farm size in Russia (for example, in Ryazan oblast it is 57 ha and in Smolensk, 54 ha). In Moscow oblast private farm size exhibits core-periphery regularity, from a few hundredths of a hectare near Moscow to 100 ha at the periphery. Private farms contribute only 1.2 percent of the oblast's agricultural output, the highest contribution being for dairy (1.8 percent of total output). Together, subsidiary and private farms in Moscow oblast occupy over 150,000 ha. (Collective orchards, gardens, dachas, and areas under individual housing construction combined occupy 177,000 ha.) Private and subsidiary farms compete directly on the market because both are "preferred users"; that is, as described in "Land Prices," below, land is assigned to them either free of charge or at normative prices by local administrations and publicly run farms.

The intensity of land use in the oblast reveals a pronounced core-periphery gradient, which also shows in the geographies of capital and labor inputs per unit of land. There is, similarly, a substantial core-periphery gradient in the number of cattle per unit of farmland. An average hectare of farmland in the innermost ring produces 10 times more meat and 3 to 4 times more milk than one in the most remote ring, and a hectare in the inner ring yields 3 times more vegetables than in the remaining rings combined. It appears, therefore, that the value of each hectare of agricultural land near the city is several times higher than on the periphery. Only southeastern districts within the flood plains of the Moscow and Oka Rivers can compete in terms of production. At a time when overall agricultural output is declining almost everywhere in Russia, such a variance takes on greater importance, especially in light of popular anxiety over the alleged loss of national food provision security because of an excessively high proportion of food imports.

Land Use: Composition and Dynamics

The spatial pattern of the central city's sprawl is fairly routine. First, corridors along major radial thoroughfares are built up, and spaces between them are gradually filled. The city's perimeter expands, while settlements growing outward from transport arteries split natural landscapes and agricultural land into small parcels that form residual green wedges between major roads extending from the central city. As the city demands more and more land for housing construction, it encroaches on those green wedges. This process began years ago within the area encircled by the Moscow Ring Road. In the period 1960–84 the Ring Road was the city boundary, and Zelenograd was the only city district lying outside the Ring. In 1984 the city obtained new land along major transport routes in the oblast. Recently, the land occupied by two of the four major airports in the area, Vnukovo and Sheremetyevo, were transferred to Moscow's jurisdiction.

Prior to the emergence of the land market in the 1990s, land-use policy in the districts of the first ring and in part of the second ring was administered by the city, although the city and Moscow oblast have always been separate administrative entities. The introduction of land prices and land transactions brought drastic changes. The relationship between the leaders (Yury Luzhkov, mayor of Moscow, and Anatoly Tyazhlov, the leader of the oblast administration) is overtly serene, but the oblast has become more assertive and is able to thwart city initiatives. Land-use policy in the closest-in districts is no longer a prerogative of the central city, and the oblast has managed to block the construction of another circumferential highway 15 km out from the current Moscow Ring Road, out of fear that it would become the new city line. Still, there is clearly some coordination between the city and oblast. Although the new ring motorway was turned down, the two will work together to reconstruct an existing bypass 60 km from the city center.

The fairly high proportion of forests in Moscow oblast, even compared with adjoining non-black-earth oblasts, may appear to be at odds with the area's high density of population and industrial activity. However, as early as the 18th century, the protective role of the woods outside the city of Moscow was recognized, and special fiats were issued prohibiting the felling of trees within 200 verstas (212 km) of Moscow. The first national park in Russia was established on Losinnyi Ostrov, which now straddles the northern border of Moscow.

Changes have taken place in the composition of the oblast's principal land users over the past 10 years. The forested area has largely been retained, but the share of state agricultural land users was reduced, gradually before 1990 and then abruptly, with the total decrease amounting to 17 percent of initial holdings. Before 1990 the residential, industrial, and transport areas of Moscow and other towns were "eating up" rural settlements and agricultural land. During the period 1950–80, for example, farmland contracted each year by 5,000–8,000 ha, or by more than the average landholding of a state farm. From 1990 on, however, it was urban areas that began to lose land. During the period 1990–96 the proportion of municipal and industrial land decreased from 13.3 to 10.8 percent. The rights of rural administrations in land management and redistribution have been substantially expanded, and these administrations have become de facto landowners. The total land under their jurisdiction increased by 245,000 ha, and their share of the total land area of the oblast increased almost sixfold, to 6.4 percent.

Although farmland occupies the largest share of land area in the most remote districts, two thirds of all losses of farmland and over 70 percent of all reductions in arable land occur in the two rings nearest the central city, especially the innermost ring. The distribution of farmland losses by geographic sector is more even, but the absolute size of farmland transfers is conspicuous not only in the vast districts of the western sector but also in the tiny districts of Krasnogorsk and Liubertsy, which adjoin the Moscow city line; these districts have lost over half their agricultural land.

Land Transfers

Unfortunately Russian land-use statistics are so convoluted that distinguishing concrete users, let alone de facto owners, is difficult without a thorough field survey. Lacking the opportunity to do such a survey for the whole oblast, we relied on the official Form 22, which is supposed to be filled out and updated by regional land-use committees. This form is a matrix whose columns are land users, such as farms of different types, municipalities, and rural administrations, and whose rows are forms of land use. Thus, land under subsidiary plots (altogether, 92,000 ha) appears in the columns for former state farms, forest administrations, and urban and rural administrations, depending on exactly where the operators of subsidiary farms live and who allotted them land. Form 22 changes from year to year as new users appear. Interpretative difficulties are manifold. For example, because most collective and state farms have nominally privatized their land, a statistical artifact emerges: the districts with the highest share of agricultural land appear to be the most advanced in terms of economic reform. Yet how can one categorize as a private enterprise a farm with 100–400 workers who have no say in management and never have had any? The current forms of shareholding may well be a transitional category of property, from collective to private, but the shareholding stage is not reflected in land-use statistics.

We therefore tried to distinguish de facto private forms of land use (private farms, subsidiary farms, orchards, vegetable gardens, dachas, and individual housing construction) from de facto collective uses, regardless of the official status of the latter. It is clear that the greatest change occurred in the early 1990s, when collective forms of land use decreased by 25 percent while private forms expanded almost fourfold. Whereas in 1985 the ratio of private to collective uses was 1 to 20, in 1996 it was 1 to 4. Of the 500,000 ha (about one third of all land within the jurisdiction of socialized farms in the late 1980s) removed from state and collective farms, over 200,000 ended up in private hands. (Actually the amount of de facto private land in Moscow oblast is even greater; for example, part of the category "other lands," which has grown substantially lately, is most likely largely private.) Owing to the expansion of land in private use, the overall reduction of farmland does not appear to be as substantial as it might seem at first glance. What has changed is not so much the amount of land as its use, which, on the smaller private parcels, has become more intensive because of the additional manual labor expended.

Like other jurisdictions of the Russian Federation, Moscow oblast tries its best to conduct its own land-use policy. Just in the period 1992–95, it adopted over 40 rulings on this subject (Oblkomzem 1996). Some of them corroborated federal laws and stipulations, while others introduced local amendments and made up for gaps in federal land-use regulations. In 1991–92 an oblast land redistribution fund was set up that contained 350,000 hectares—the excess of the area of state land to be privatized over the total amount of assigned land shares. Three quarters of this amount was transferred to the jurisdiction of rural and small-town administrations; districts distributed the remaining quarter

(Oblkomzem 1996: 7). Under a September 1992 ruling of the Moscow oblast soviet, the decision to transfer land within the confines of rural or urban soviets (later renamed administrations) belongs to the soviets; the district has the power to dispose of land outside designated settlements. Finally, important land uses, such as national parks and defense-related areas, are administered at the oblast level.

Citizens and institutions apply for land to the appropriate authorities, which assign the requests to their land-use committees for decision. Such decisions are supposed to be coordinated with building permit, sewage treatment, and fire inspection authorities and with current land users whose interests are affected by potential land transfers. Finally, the decision must be endorsed by the local administration. Another ruling of the oblast administration defined sizes of land parcels. For example, a private farm may not exceed 40 ha or be less than 2 ha, while individual parcels in collective orchards can be from 0.02 to 0.12 ha.

The above rulings were introduced during the period of administrative redistribution of land, before the law on land prices was adopted, so they do not stipulate any financial transactions. Now every citizen has a right to buy land. Waiting lists for free parcels, a vestige of the old ways, still exist in some districts, but today almost everything hinges on purchase offers. However, the lack of a Federal Land Code, the attendant uncertainty about rights subsequently to sell newly acquired land, and the secret character of land deals beget disarray and disorder in land-use allocation. This disorder is more than tainted by corruption and profiteering. In most cases the general public does not know how local authorities spend the money made from selling land that was acquired virtually free of charge from state farms and other users in the early 1990s.

The authorities seem to be helpless in the face of the semicriminal character of the land market. According to reports in the media, the Moscow city and oblast governments are increasingly concerned about private construction projects going forward without permits. Proposals to bar electricity and water services for these unlawful estates have been made. However, people who can afford to bribe local officials, grab incredibly valuable land parcels, and plug into the local infrastructure can also afford to skirt all other obstacles.

Land Prices

Two kinds of prices have to be distinguished: *normative,* that is, set by the oblast administration, and *market,* controlled by supply and demand. In the wake of the Federal Law on Payment for Land, in June 1995 a local ruling, "Payment for Land in Moscow Oblast," was adopted that set the tax rate on each type of oblast land except farms at 2.5 times the Russian average for that type of land.[8] For farmland, 10 classes were designated on the basis of soil type and improvements. Tax rates for these classes were further differentiated according to location with respect to the city of Moscow. In all cases the normative price was set as the land tax times 200. According to a later federal ruling, these prices were to be adjusted annually, and the first such adjustment took place in 1996. The highest-quality agricultural lands in the immediate environs of Moscow were priced at $680 per ha, while the worst and most remote farmland was priced at $76 per ha. Within accessibility rings, prices vary by a factor of 2.6 according to quality; for land of the same designated quality, prices vary by a factor of 3.3 on the basis of accessibility. Thus, normative pricing recognizes the preeminence of location as a price factor, a huge departure from the Soviet-style economic mentality. There is only one clear exception to the concentric pattern of prices: a wedge of valuable soils in the southeast within the fertile flood plains of the Moscow and Oka Rivers.

Even after adjustment, normative prices remain far below spontaneously set market prices. A comparison of normative and actual market prices by Filippovich shows that the price differential amounts to a factor of 1,000 in some districts of the inner ring and that the market price is "only" 50 times the normative one within the confines of settlements. The geography of market prices clearly emphasizes location much more than the physical quality of land. It should be noted that the market prices under discussion concern land parcels for second, recreational dwellings, the only area in which buying and selling has occurred more or less in the open.

Market prices have not changed much since 1995, but some changes did take place between 1993 and 1995. Whereas in early 1993 land in the innermost ring was 17 times more expensive than in the outermost one, by 1995 the difference was 80 times. In the prestigious vicinity of Nikolina Gora (at the juncture of Odintsovo and Krasnogorsk rayons) one sotka (0.01 ha) without any built structures cost over $4,000, but in Taldom, in the far north of the oblast, one sotka cost less than $100.

Prices for smaller (0.05–0.07 ha) land parcels in collective orchards are somewhat lower than for dachas, but the spatial pattern is roughly

---

[8] The federal law was adopted in October 1991 but was not enforced until November 1995, when the federal government issued its ruling on land prices and taxes. For the Moscow oblast ruling, see Oblkomzem (1996): 154–60.

the same. In May 1996, 0.01 ha cost over $1,000 within a 20-km radius of Moscow but $70 at a 150-km radius.[9] The availability of some infrastructure may boost prices dramatically, by 20 to 30 percent near the Moscow Ring Road and by as much as 60 percent in distant areas. (Land prices do not include the prices of structures.)

Transacting legal recreational land deals is fairly cumbersome, but doable, and people often accomplish it without go-betweens. Most conflicts are due to claims made by the family of a seller. Collective orchard comradeships may impose bans on selling land to nonmembers or may stipulate certain procedures.

Our March 1997 field observations of recent land-use changes in two districts, Krasnogorsk, immediately west of Moscow, and Shchiolkovo, northeast of the city, give concrete expression to the above generalities. Our observations confirmed the crucial role of localities and district administrations in land reallocation processes. A characteristic solution resorted to in both exurban districts was to purchase land for local residents' agrorecreational dwellings in more remote districts of the oblast in order to be able to sell local land to wealthier customers from the city of Moscow.

Conclusions

We began this analysis by outlining the crucial differences that have set Russian urban margins apart from their Western, especially American, counterparts. These differences are so significant that they have not changed dramatically during the current stage of transition, although some developments, such as cottage construction, have already begun to "Americanize" Russian suburbs. Like other historical borrowings from the West, on Russian soil the phenomenon has all but shed its original substance (suburbia as a middle-class spatial niche) while retaining its original form (a detached family house outside a city). Still other trends, such as squeezing out agriculture from the first accessibility ring, are similar to what happens in the West.

Perhaps the most important feature of the Moscow Urban Field is the burgeoning land market generated by demand for second dwellings. The condition of this market underlies, although in large measure informally, the financial well-being of territories and their populations. This market has a semihidden character, most of its inner workings being arduously kept away from observers and analysts. The nascent land

---

[9] "Skolko seichas stoyat shest sotok," *Izvestia,* May 9, 1996.

market in the Moscow urban area, which is bursting through administrative restrictions and outdated preconceptions, is unique in Russia in its magnitude and its ability to influence daily life. Only in the vicinity of St. Petersburg can one find something similar.

Our analysis has shown that the environs of large Russian cities are still important magnets for migration. Recently, the Constitutional Court of the Russian Federation took up the constitutionality of the Moscow oblast law mandating that incoming migrants pay for their registration as permanent residents of the oblast "to compensate costs of developing infrastructure," as the title of the law puts it. The regulation requires that every migrant deposit the equivalent of a minimum of 300 monthly salaries, which in June 1997 amounted to 25 million rubles ($4,310)—a huge sum for ordinary Russians.[10] (Until recently, a migrant coming into the city of Moscow was required to deposit the equivalent of 500 minimal monthly salaries.) The very existence of such stipulations, unknown in democratic countries, clearly testifies to the heightened attractiveness not only of large cities themselves but of their environs as well. The fee filters prospective migrants, retaining only the wealthy in the pool. It is a glaring example of the power of increasingly independent regional authorities—a power that shows up in other spheres of life such as taxation, subsidies, privatization schemes, and regional protectionism. As Latynina (1997) noted in another context, "In Russia, the breakup of the last traditional empire begot a quasi-market order, under which it isn't goods that are the principal article of commerce but tax favoritism, government rulings, and access to extrabudget funds." Our research shows that although land itself is not quite privatized, the right to distribute it is, and local authorities have thus become almost the main beneficiaries of land sales.

What Latynina calls "financial feudalism" has not precluded some developments in the vicinities of Russia's large cities that are similar to those experienced earlier by many economically advanced countries—for example, the declining role of "suburban agriculture." However, in Russia, where core-periphery gradients in agricultural land use do not hinge on transport costs as much as on the depopulation of the vast rural periphery, the decline of farming in the "city's countryside" may be a mixed blessing, at best. A sensible course for Russian suburbia may be the introduction of farmland preservation strategies of the kind that exist in Montgomery County, Maryland, which include "buying easements, or restrictions, on thousands of acres of farmland as a fire wall against

---

[10] "Sbory mogut i otmenit," *Izvestia,* June 18, 1997.

the spread of housing tracts."[11] In other words, farmers would continue working their land in exchange for selling developers the "rights" to build homes elsewhere in the environs of a large city. Zoning ordinances may have to be enacted, as well.

These preservation strategies seem to be unavoidable because of the high degree of spatial coincidence of the demands for recreational and agricultural land. Because the Russian notion of a good dacha implies a much bigger plot than a single-family detached house in America usually has, the appetite for land in Russia is potentially greater. This appetite may affect the development of the environs of Moscow, St. Petersburg, Novosibirsk, Ekaterinburg, Perm, and the few other very large cities of Russia. The prospects for a heated competition for land around smaller provincial centers in the near future are low.

The environs of Russian cities—areas that are neither quite urban nor rural—have not attracted much attention among specialists in Russian area studies. It is high time to fill this void. Urban margins are unique in terms of the sheer magnitude of the ongoing land-use shifts, and land use, and its attendant regulations are an important aspect of Russia's economic and political reforms. Although it is not yet clear whether changes in land use will be truly profound and far-reaching all across Russia, exurbia is already a testing ground for those changes; only there, so far, has a free market in land, at least in some instances, become part of public life. Whatever pattern suburban land use in Russia assumes, it needs to be carefully monitored, as it speaks volumes about the real state of the Russian economy.

References

Barmina, I. 1995. "Sotka za sotku." *Argumenty i Facty,* October 16–22.
Bryant, C. R., and T. R. R. Johnston. 1992. *Agriculture in the City's Countryside.* Toronto, Canada: University of Toronto Press.
Faruseth, O. J., and J. T. Pierce. 1982. *Agricultural Land in an Urban Society.* Washington, D.C.: AAG.
Filippovich, L. "Gorozhanie."
Fehr, Stephen C. 1997. "Montgomery's Line of Defense against the Suburban Invasion." *Washington Post,* March 25.
Fishman, Robert. 1987. "American Suburbs/English Suburbs: A Transatlantic Comparison." *Journal of Urban History* 13 (3): 237–51.
Goskomstat (State Statistical Commission). 1995. *Demograficheskiy yezhegodnik Rossiyskoi Federatsii.* Moscow.
———. 1996. Rossiyskiy statisticheskiy yezhegodnik 1995. Moscow.

---

[11] Fehr (1997); see also Faruseth and Pierce (1982): 60–66.

Ioffe, G., and T. Nefedova. 1997. *Continuity and Change in Rural Russia.* Boulder, Colo.: Westview Press.
Jackson, Kenneth T. 1985. *Crabgrass Frontiers.* New York: Oxford University Press.
Lappo, G. M., and P. M. Polyan. 1996. "Novye tendentsii v izmenenii geourbanisticheskoi situatsii v Rossii." *Izvestiya AN SSSR, Seriya Geograficheskaya,* no. 6: 11.
Latynina, Y. 1997. "Rossiya dvizhetsya vperiod, no v obratnom napravlenii." *Izvestia,* June 17.
Logos. 1996. Rossiyskiy statisticheskiy yezhegodnik 1996. Moscow.
MFGO. 1989. Problemy uluchsheniya ekologicheskoy situatsii i ratdionalnogo priridopolzovaniya b Moskovskom Regionie. Moscow.
Oblkomzem. 1996. Regulirovaniye zemelnykh otnosheniy na territorii Moskovskoy Oblasti. Moscow.
Treivish, A. 1996. "The Central Economic Region and Moscow Oblast." In M. J. Bradshaw and J. B. D. Shaw, eds., *Regional Problems during Economic Transition in Russia: Case Studies.* TASIC, Russian Regional Research Group Working Papers Series 1. University of Birmingham.
Zayets, E. S. 1993. "Analiz novykh podkhodov k razvitiyu prigorodnoi zony Moskvy." Problemy zemlepolzovaniya v svyazi s razvitiyem maloetazhnogo zhilishchnogo stroitelstva v Moskovskom Regionie. Moscow: Mosoblsovet.

Adaptation and Integration of Forced Migrants in Russia[*]

Galina Vitkovskaya

With the first symptoms of the disintegration of the U.S.S.R. in the late 1980s, the scale and character of migration to Russia from the union republics changed abruptly.[1] During the first 5 years after disintegration (1992–96), net migration from the new independent states to Russia was 2.9 million; the total for the preceding 15 years was 2.4 million (Vishnevsky 1998: 109).

Resettlement in the post-Soviet years has been largely involuntary. Whole ethnic groups were ousted from the territories in which they lived (Armenians from Azerbaijan, Azerbaijanis from Armenia, and Meskheti Turks from Uzbekistan). Other ethnic minorities left their places of residence, mostly to return to their "native" territories. The political changes especially affected Russian-speaking people living in non-Slav countries. In the migration exchange with Russia after 1991, the non-Slav countries lost 17 percent of their Russian populations, whereas the Slav countries lost only 2 percent (Vishnevsky 1998: 113).

The Survey

In 1997–98, with support from the International Organization for Migration, the author conducted a survey in five *oblasts* (regions) in European Russia: Orel, Rostov, Ryazan, Saratov, and Voronezh. The selected oblasts differ as to area, population, urbanization, migration load, policies regarding refugee status and residence permits, and social and economic conditions at the time of the survey (see Table 1). They are, by and large, "normal," without extreme social tensions or odious policies toward resettlers (as practiced, for instance, in Moscow). The sample allows an examination of average trends in migrants' adaptation in various types of settlement. Earlier regular surveys and a study of Tver oblast in late 1998 provide additional data for comparison of developments over time.

---

[*] Reprinted with permission of International Office on Migration, Geneva.
[1] Forced migration flows prior to the end of 1991 were usually associated with conflicts. These bloody conflicts of 1988–90, in Sumgait, Nagorno-Karabakh, Baku, Fergana, Uzgen, Osh, Bishkek, the Naryn and Talas regions of Kirghizia, Dushanbe, Vilnius, Tbilisi, and so on, were symptoms of growing separatism and the dramatic weakening of the center.

Table 1. General Characteristics of the Oblasts Covered by the Survey

Characteristic	All Russia	Oblast				
		Orel	Ryazan	Voronezh	Rostov	Saratov
Area (1,000 square kilometers)		24.7	39.6	52.4	100.8	100.2
Number of towns under oblast jurisdiction		3	4	7	16	12
Number of towns under rayon (district) jurisdiction		4	8	8	7	5
Number of urban-type settlements		14	27	22	25	36
Number of rural area administrations		222	484	491	441	591
Share of urban population as of January 1, 1998 (percent)	73	63	68	62	68	73
Population of oblast center as of January 1, 1998 (1,000 people)		347.6	539.4	982.7	1,024.5	898.7
Population as of January 1, 1998 (1,000 people)	146,739	905.8	1,308.4	2,482.5	4,387.6	2,721.4
Number of those granted refugee or forced migrant status, per 1,000 inhabitants, as of January 1, 1998	8.1	13.3	8.2	13.5	8.7	18.6
Migration increase due to the inflow from the new independent states, 1997 (number of people)	433,368	3,386	4,324	8,902	11,031	13,035
Migration increase due to the inflow from the new independent states, 1997 (number of migrants per 1,000 inhabitants)	3.0	3.7	3.3	3.6	2.5	4.8
Ratio of official forced migration load to total migration due to the inflow from the new independent states, 1997[a]	0.3	0.25	0.29	0.18	0.21	0.25
Ethnic composition of the population (percent)						
Russians		97.0	96.1	93.4	89.6	85.6
Ukrainians		1.3	1.2	5.0	4.2	3.8
Armenians					1.5	
Kazakhs						2.7
Tatars						2.0
Money income per capita, 1997 (1,000 rubles)	821.8	605.1	505.5	551.0	466.2	443.7
Subsistence level, 1997 (1,000 rubles)	370.7	259.6	282.9	290.2	265.4	333.9
Ratio of money income to subsistence level (percent)	222	233	179	190	176	133
Official registered unemployment, 1997 (percentage of labor force)	3.4	2.0	2.3	2.5	1.2	3.4
Volume of industrial production in 1997 as percentage of 1991 level	52	36	33	35	34	41

a. An indirect indicator of the policy regarding the granting of an official status.

Sources: Goskomstat (1998): 11–12, 44–45, 98–99; Moscow Center of the Carnegie Endowment for International Peace (1998a), vol. 2: 505, 754, 795, 809, 835; data from oblast committees on state statistics.

During the period October 1997–April 1998, a random sample of 888 forced migrants and 771 local residents was polled. The migrants were from the former union republics of the U.S.S.R. and from conflict zones within Russia, mainly Chechnya. The survey covered rural areas (villages), small towns and urban-type settlements (referred to as "small towns" in the text), and regional centers (designated "large and medium-size cities" or "cities" in the text).

Characteristics of the Recent Migration

Ordinarily, migrants move after and because they have found an attractive job; they do not go to places where their employment prospects are worse. Forced migrants, by contrast, move because job discrimination and economic conditions have created an unfavorable employment situation in their places of residence and because they see no prospects for themselves in the new geopolitical conditions. They regard permanent resettlement, rather than temporary moves or trips to earn extra money, as their only option.

The educational and vocational level of forced migrants is very high: on average, it noticeably exceeds that of the host population in Russia. According to the 1989 population census, the Russian population in Russia had, on average, 8.8 years of education, as against 9.6–10.1 years for Russians in the Transcaucasian republics, 9.0–9.5 years for those in Central Asia, and 9.3–9.6 years for those in the Baltic republics. The share of people with higher education among the population of Russia was 11.3 percent in 1989, whereas for the forced migrants it was 19.5 percent in 1997.[2] Their qualifications make migrants serious competitors in the struggle for jobs.

Settlement Patterns and Adaptation

Migrants' settlement patterns in Russia are poorly matched with their urban origins and social composition. There are several reasons for this.

First, forced migrants had little opportunity to prepare for their move. Only 39 percent of those questioned had made advance preparations; 59 percent left unexpectedly (34 percent after some particular event, 21 percent because of force or out of fear for their lives, and 4 percent as evacuees from a war zone). As a result of the growing separation of the countries of the former U.S.S.R., a sharp rise in prices of

---

[2] The figure for the Russian population is from Finansy i Statistika Publishers (1990): 6; for forced migrants, the source is Goskomstat (1998): 114.

conveyance, a decline in income levels, and problems with selling property, potential migrants have great difficulty in obtaining information, making reconnaissance trips, and paying for the move. Second, the housing situation and the policies regarding residential registration, migrants' status, and assistance compel forced migrants to settle in small towns and villages—exactly the places where they are least likely to adapt well.

In our study, two main criteria were used to evaluate the success of adaptation by forced migrants: the migrants' general assessment of their current situation, and their intentions as to further migration. Complex indicators of the success of adaptation were constructed that included these factors and the migrants' responses as to what they liked and disliked about their new places of residence. The analysis shows that the highest probability of successful adaptation and the lowest probability of poor adaptation are found in large and middle-size cities (Figure 1).

On average, 57 percent of those interviewed did not want to leave their current place of residence in Russia, but 23 percent did want to leave—8 percent, as soon as possible. Migrants in large and medium-size cities were most inclined to stay; those in rural areas were most likely to want to leave.

Figure 1. Adaptation of Migrants in Various Types of Settlement

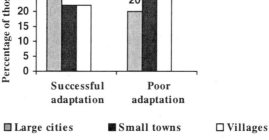

The probability of successful adaptation grows with each year in the new place and after five years is considerably higher than the average. Conversely, the probability of poor adaptation gradually decreases. This happens not only because of real adaptation of the forced migrants but also because those who have not adapted well gradually leave their original place of resettlement, increasing the proportion of successful adapters in that place.

A comparison of adaptation outcomes by type of settlement demonstrates the categorical advantage of settling in cities, for all social and demographic groups (Table 2). For the sample as a whole, 25 percent adapted successfully, and 31 percent showed poor adaptation. All age groups have a lower than average probability of successful adaptation in villages and a much higher than average probability of unsuccessful adaptation. Men find it much harder to adapt in villages, and women have more difficulty in small towns. For the educational group with the most successful overall adaptation (that is, those at a low educational level), the probability of successful adaptation in cities is very high—nearly 70 percent—but it falls below the average in small towns and is far below the average in villages.

No matter how favorable the characteristics of the migrants themselves, their adaptation depends predominantly on the objective situation they find themselves in after moving. Three groups of factors are important:

1. *The situation in the new place of settlement.* This includes the "comfort factors," such as relief from wars, conflicts, threats, and discrimination; ethnocultural and ethnosocial compatibility; and a feeling of return to the motherland. The importance of the comfort factors for adaptation depends greatly on other facets of the situation in the new place of residence. Two factors related to poor adaptation are (a) the unfamiliar circumstances of life in small towns and villages and (b) bad environmental conditions in small towns.

2. *Accommodation of migrants:* housing, employment, income, and daily living conditions.

3. *The attitude of the receiving social community*—both local people and authorities—toward the arrival of forced migrants.

Table 2. Adaptation of Groups with Different Social and Demographic Characteristics, by Type of Settlement

(percentage of those interviewed; numbers in parentheses are deviations from the average proportion for the total)

Characteristic	Successful adaptation				Poor adaptation			
	Large and medium-size cities	Small towns		Villages		Large and medium-size cities	Small towns		Villages	
Age										
Youths (age 18–29)	25 (0)	16	(–9)	14	(–9)	17	(–14)	32	(+1)	48	(+17)
Middle and older employable age group	28 (+3)	21	(–4)	24	(–1)	25	(–6)	34	(+3)	42	(+11)
Pension age group	44 (+19)	36	(+14)	21	(–4)	11	(–20)	31	(0)	35	(+4)
Gender										
Male	30 (+5)	27	(+2)	19	(–6)	20	(–11)	28	(–3)	52	(+21)
Female	31 (+6)	19	(–6)	24	(–1)	20	(–11)	36	(+5)	34	(+3)
Education										
Higher	26 (+1)	24	(–1)	22	(–3)	28	(–3)	33	(+2)	43	(+12)
Higher incomplete	22 (–4)	13	(–12)	21	(–4)	39	(+8)	47	(+16)	64	(+33)
Secondary special	28 (+3)	24	(–1)	24	(–1)	16	(–15)	40	(+9)	43	(+12)
Secondary general	33 (+8)	21	(–4)	24	(–1)	14	(–17)	16	(–15)	34	(+3)
Secondary incomplete; primary	67 (+42)	21	(–4)	7	(–18)	—	(—)	25	(–6)	21	(–10)

— Not applicable (no members of group in the sample).

Employment

In our surveys of 1991, 1992, and 1993, the availability of housing was decisive for the adaptation of forced migrants. In the 1997–98 survey, employment was more crucial than housing (Table 3). Its heightened importance may be partly related to a gradual change in the social mentality, to an erosion of the paternalistic frame of mind. It is consistent for forced migrants to be pioneers in such changes, for they are struggling to survive in extreme conditions. To solve their housing problem, they look not so much to the state as to their own efforts. But to obtain housing for themselves, they need jobs, and employment becomes prominent as a priority.

Table 3. Effect of Various Aspects of Migrants' Situation on Adaptation

Characteristic being surveyed: dissatisfaction and problems	*Number of persons*	*Effect of the characteristic (percent)*	
		On successful adaptation	*On poor adaptation*
What don't you like in the new place of residence?			
Unfamiliar conditions of life	23	—	+34
Bad ecology	32	–6	+31
Housing problems	88	–10	–3
Employment problems	106	–9	+15
Bad daily living conditions	81	–4	+7
Lack of roads and infrastructure	42	–6	+7
Economic situation; low living standards	30	–2	+19
No prospects	18	–14	+13
Attitude of local people	97	–7	+12
Attitude of local authorities	41	–15	+13
What are the most vital problems for the family?			
Ecology, climate	69	0	+11
Employment	443	–9	+9
Housing	518	–8	+4
Residence registration	45	–12	–7
Economic situation	700	–3	+1
Attitude of local people	16	0	+7
Attitude of local authorities	13	–2	—

a. Defined as the deviation of the proportion of the group with successful or unsuccessful adaptation (in the group with the specific characteristic) from the average proportion of the group with successful or unsuccessful adaptation among all those interviewed.

The employment situation of forced migrants in Russia is sharply different from that before their move. Before migration, nobody in the sample lived from odd jobs, but in the new places of residence such employment has become the only source of subsistence for many of them. Unemployment (excluding odd jobs) among forced migrants was 7.9 percent in the fourth quarter of 1997 and 8.3 percent in the first quarter of 1998; for the Russian population as a whole, it was 3.4 percent in 1997.

During the first year after arrival in the new place of residence, migrants in small towns find a better employment situation, but after five years in one place, the situation in small towns is much worse than in other types of settlement (Figures 2 and 3). Villages have the highest level of permanent employment, and cities have the lowest level of unemployment.

Skills and Vocational Status

Only about 40 percent of the employed forced migrants work in the same branch in which they were engaged before moving. The share of those working in industry and science is sharply reduced, particularly in small towns and villages (Table 4), and the share of those employed in agriculture is substantially higher. Education and public health workers are more likely to succeed in preserving their specialization.

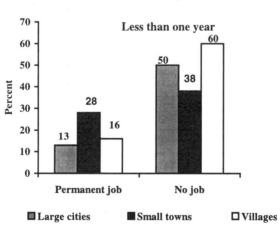

Figure 2. Employment of Economically Active Resettlers, by Length of Residence

Table 4. Occupation by Sector before and after Moving, by Type of Settlement

(percentage of employed forced migrants residing from one to five years in a new place)

Sector	Large and medium-size cities		Small towns		Villages	
	Before moving	After moving	Before moving	After moving	Before moving	After moving
Industry	32	22	22	8	23	8
Agriculture	3	4	8	11	10	29
Construction	11	9	14	13	6	6
Transportation and telecommunications	4	4	5	4	6	8
Science	4	1	3	—	5	—
Education	8	8	14	8	18	15
Culture and sports	2	3	2	3	2	1
Public health	5	4	4	4	1	1
Trade	5	16	6	19	11	13
Services	4	7	7	5	1	3
Government	11	8	9	12	10	5
Other	10	14	6	13	7	11

The vocational and qualification status of approximately one third of the employed forced migrants who have been living in a new place of residence between one and five years is lower than their status just before moving. This is especially so for migrants living in villages (Figure 4). Only 14 percent of the "old-time" migrants living in large cities for more than five years said that they would never be able to regain their former social status, but 23 percent of those living in small towns and 29 percent of those living in villages gave that response.

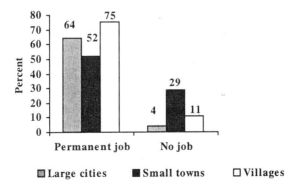

Figure 3. Employment of Economically Active Resettlers, by Length of Residence
More than five years

Entrepreneurship

Migrants have set up numerous industrial, educational, publishing, and building enterprises, and self-employment is widespread among them. Substantial assistance with financing and equipment has been rendered by international organizations, notably the International Organization for Migration (IOM) and the United Nations High Commissioner for Refugees. Nongovernmental agencies for the support of migrants have also emerged, on both the regional and the national levels. The biggest of these, the Compatriots Foundation, assists migrants' organizations in such spheres as marketing, consulting, and leasing.

It is on entrepreneurship that forced migrants are pinning many of their hopes for settling down in Russia and regaining the material and social status which they lost after moving. Our surveys of 1991, 1992, and 1993 indicated that the potential entrepreneurial activity of forced migrants was higher in the countryside than in cities.

Tax policy was mentioned as an obstacle to entrepreneurship by 14 percent of those questioned; red tape, by 10 percent; corruption, by 10 percent; and a disapproving attitude on the part of authorities, by 1 percent (4 percent in villages). Seven percent (5 percent in cities and 11 percent in villages) noted that all areas of activity were already controlled by someone. Opposition by authorities is stronger in the countryside, but when authorities are supportive, it is easier for migrants to build up their businesses in the countryside than in the city.

Housing and Wealth

Russians living in the former union republics (especially in Central Asia) occupied mostly very good and quite often elite housing. Of the

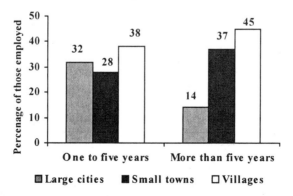

Figure 4. Share of Migrants Suffering Decline in Vocational Status, by Type of Settlement and Length of Residence

total number of forced migrants questioned in 1997–98, 22 percent had their own houses before the move, and 71 percent had their own (privatized) apartments. It was, in the first place, their housing that they lost as a result of migration. When asked what difficulties they had encountered in moving, 40 percent of the respondents replied that it was "impossible to sell or exchange housing," and 2 percent said that their housing had been destroyed or left behind when they fled from a war zone. Only about 15 percent of the respondents were able to purchase or build housing in Russia; 58 percent identified the purchase of housing as one of the three problems that were the most essential for their families to solve.

At an intermediate stage of adaptation (one to five years' residence), resettlers who live in cities are in the worst situation in terms of housing: only 19 percent have self-contained housing units (houses or apartments). The figure for small towns is 36 percent and for villages, 45 percent. However, the accommodation of forced migrants in employer-owned housing, which is especially prevalent in villages (see Table 5), substantially reduces the probability of successful adaptation and increases the probability of poor adaptation.

Table 5. Housing of Forced Migrants, by Type of Settlement and Length of Residence

Type and ownership of housing	Large and medium-size cities		Small towns		Villages	
	1-5 years	>5 years	1-5 years	>5 years	1-5 years	>5 years
Self-contained private or municipal housing units	14	51	27	68	22	24
Private	14	38	21	55	20	9
Municipal	—	13	6	13	2	15
Employer-owned housing for temporary or permanent residence	5	8	9	13	23	20
Municipal shared housing units	—	—	1	—	1	—
Employer-owned shared housing units for temporary or permanent residence	2	3	1	—	5	—
Room in a dormitory	38	28	25	2	20	28
Bed in a dormitory	6	—	1	—	1	—
Housing rented from private individuals	17	5	16	10	10	15
Relatives' or friends' dwellings	13	5	12	5	10	6
Barracks (private, municipal, or employer owned)	1	—	5	2	7	7
Other lodgings poorly suited for living	2	—	3	—	1	—

— Not applicable (no members of group in the sample).

Market prices for housing in small towns (except in towns located close to Moscow or St. Petersburg and in such places as resorts) are the lowest in Russia, partly because of the lack of employment. Migrants who bought apartments or houses in these towns face low demand for their housing and cannot sell out to move where labor market conditions are better.

Migrants in villages have larger housing, on average, than those in urban settlements, but their housing situation does not improve much over time, and after five years, it is worse than for migrants in cities. The main reason is the small share of privately owned housing, which is generally roomier than housing belonging to other owners and, particularly, than municipal housing.

Forced migrants lost considerable property on moving. A survey conducted by the Moscow Carnegie Center in Tver oblast in late 1998 shows that migrants had been rather well to do.[3] About one third of them had automobiles, whereas only 24 percent of the population of St. Petersburg did (Moscow Center of the Carnegie Endowment for International Peace 1998b: 278).

Forty percent of the resettlers said that they had used all their funds to pay for their move—which is quite in keeping with the realities, considering the dumping prices for the property they had to sell and the bribes and charges expended in moving and in transporting luggage. Only 8 percent of migrants have bought housing in Russia with the money they brought with them; 6 percent have bought a plot of land.

The resettlers who were most successful in solving their property problems tend to go where they can buy housing at a relatively low price—to villages and small towns. Those who were able to take along nothing or only clothing prefer cities, which offer the best prospects for finding a good job with good pay; 31 percent settled in large cities, 26 percent in villages, and 21 percent in small towns.

A low level of pay and lack of property are behind the rather low overall wealth of the families of forced migrants. What stands out particularly is the share of starving families living in villages.[4]

---

[3] The survey was conducted within the framework of the project "Migration and Safety" being implemented by the migration and citizenship program of the Moscow Carnegie Center. A random sample of 152 local residents and 100 forced migrants in Tver oblast was surveyed. The questionnaire included a number of questions asked in previous surveys, in the same wording.

[4] Among the reasons for the unhappy lot that can fall to migrants in villages are low pay, difficulties with finding jobs, problems with acquiring a garden plot, the resettlers' urban

Relations with the Local Population

Adaptation to a new social setting is often the most complicated element in the general process of forced migrants' adaptation in their new places of residence, and also the most painful. Migrants' perception of an unfriendly or hostile attitude toward them in their new social setting dramatically increases the probability of poor adaptation.

Antimigrant Moods

As of January 1, 1998, according to official registration data, Russians constituted 76 percent of forced migrants; Tatars, 7 percent; and members of other peoples of the Volga area (Bashkirs, Mari, Mordovians, Udmurts, and Chuvash), 2 percent.[5] Migrants are not always given a friendly reception in the mother country.

The current process of forced migration has no precedents in Russian history. Although repatriation of Russians began long ago—in the 1960s in Transcaucasia and in the 1970s in Central Asia—xenophobia among the Russian population toward migrants was not observed (or was not regarded as a problem) in those years. The earlier migration differed from the current one in several ways:

- It was considerably smaller in scale.

- Resettlement occurred on an individual basis and was widely scattered.

- The situation in Russia's labor markets was much more favorable than in the countries of exit or in Russia today.

- Repatriation was predominantly voluntary. The motives for moving were positive, and the migrants were attracted to their new homes rather than driven out of the old. The places

---

background and lack of skills essential for effective management of a farm, and the radically different climatic zone, which requires mastery of new agrarian technologies.
[5] Russians go mainly to "Russian" regions; Tatars, Bashkirs, and members of other Volga-area peoples generally return to their traditional settlement areas (the ethnic republics of the Volga area and the southern Urals, where Russians prefer not to go). In 1996, 80 percent of the Bashkirs who had been granted refugee or forced migrant status moved from the former union republics to the Ural region of Russia. Fifty-nine percent of them settled in Bashkortostan and nearly all the rest in the Volga area. The probability of Bashkir migrants' settlement in the Ural region is 5.5 times the probability of the overall inflow of forced migrants into this region. The probability of Tatar resettlers' settlement in the Volga area is 2.1 times the overall figure for migrants; in the Ural region it is 2.6 times. The probability of Russians' settlement in Udmurtia is half the overall probability of the inflow of forced migrants there, in Tatarstan it is less than one third and in Kalmykia, less than 1/30 (Vitkovskaya 1998: 36–38).

of residence chosen were appropriate to migrants' social composition and their level of urbanization; resettlers found themselves in a suitable sociocultural setting.

All these factors have changed in the past decade. The most important differences that may have influenced public mentality and increased the level of migrantophobia are:

- *Deterioration of labor markets* accompanied by unemployment, social stratification, and widespread impoverishment. These factors increase worries about additional competition.

- *Regionalization, and politicization of the idea of regionalism,* resulting in mobilization of local identities that previously existed in passive, latent forms and in a sharpening of these identities on the basis of opposition to "aliens." The swift decrease in the territorial mobility of the Russian population after 1992 contributed to this process, as it substantially limited contacts and strengthened the isolation of local communities.

- The *mass-movement, forced character* of the flows of repatriates to Russia.

Although phobia against forced migrants in Russia cannot be described as a mass or dominating phenomenon, it is, unfortunately, noticeable.

Local Residents' Attitudes

Twenty-two percent of the local residents we surveyed had a rather negative (17 percent) or openly negative (5 percent) attitude toward resettlers' taking up residence in their areas. In four of the five surveyed regions, 25 to 27 percent of the population expressed a negative attitude. (In Orel oblast the figure was 12 percent.) These assessments do not include answers that imply a negative attitude toward particular ethnic groups.

In the survey conducted in Tver oblast in November–December 1998, 44 percent of local residents expressed rather negative (30 percent) or openly negative (14 percent) attitudes toward the inflow of forced migrants; the openly negative group included 3 percent extremely negative. It may be assumed that the increased migrantophobia is linked to the effects of the August 1998 crisis, which sharply aggravated the labor market situation and brought about a decline in the living standards of the population. Tolerance of resettlers may have decreased elsewhere, as well, particularly in regions such as

Saratov that have high migration loads. The tolerance index (the ratio of positive to negative assessments) of the Russian population toward forced migrants was 1.4 in the first survey and 0.64 in the second (Tver) survey. Thus, in late 1997 and early 1998 the population's tolerance index for resettlers was the same as for Azerbaijanis in 1997 (1.4); in late 1998 in Tver, however, it was lower than that for Chechens (0.76).[6]

Resettlers cannot help but feel the mood of the population. Of the forced migrants questioned in the 1997–98 survey, 13 percent assessed local residents' attitudes as unfriendly (11 percent) or hostile (2 percent), with not much regional variation. The share of negative responses does not include such answers as "indifferent" (35 percent) that have a negative connotation. In late 1998, 14 percent of resettlers who had taken up residence in Tver oblast sensed an unfriendly attitude. In a survey conducted by Vitkovskaya in Volgograd oblast in 1993, 12 percent of forced migrants' answers about local attitudes indicated perceived unfriendliness. Thus, the share remains rather stable in different surveys.

Local residents were asked what negative aspects they saw in the arrival of forced migrants. The expressions of discontent may be divided into two groups: a view of migrants as competitors, and a perception of their negative qualities.

The first group includes answers reflecting fears about additional mouths claiming shares of the small Russian pie. Tied for first place in this group were responses concerning jobs and housing. The third top response was, "They are wealthier than we are, and yet they are asking for aid."

In both surveys the number of answers that had to do with competition was much greater than the number that mentioned negative qualities of resettlers (2.0 times as large in 1997–98 and 2.5 times as large in late 1998). This gives grounds for regarding migrantophobia as something generated by the difficult economic situation of the Russian population rather than by xenophobic reactions and stereotypes.

In the first survey the local population was more often negative about resettlers' skills or desire to work (11 percent) than about their behavior (3 percent). Of the total number of local residents who negatively assessed forced migrants as workers, 58 percent did not work in the same places as resettlers; thus, their assessments were largely based on popular notions.

---

[6] All-Russian Center for Public Opinion (VTsIOM).

Negative attitudes toward forced migrants were expressed by 16 percent of the local residents questioned in cities, 25 percent in small towns, and 20 percent in villages. In the countryside, local residents are much more likely to see something negative in the arrival of resettlers and they are less likely to give the answer "nothing negative" (Table 6). Whereas 19 percent of local respondents in large cities see nothing positive in the inflow, this answer is given by more than 30 percent of those in villages and small towns.

Local people in small towns and villages are considerably more likely than city people to notice that forced resettlers are living in the locality. A negative attitude toward forced migrants as competitors for scarce benefits is observed far less often in cities, where the labor market situation is a little better than in villages and depressed small towns. Lower awareness manifests itself in less categorical judgments; residents of large cities are much more likely to find it difficult to give a concrete answer about the negative consequences of the inflow of migrants. In villages and small towns the differences in the degree of negative response are to a great extent determined by the actual strength of competition for housing and jobs (especially in small towns) and for plots of land (in the countryside).

Table 6. Negative Aspects of the Inflow of Forced Migrants according to Local Residents, by Type of Settlement

(percentage of local residents questioned)

Response	Large and medium-size cities	Small towns	Villages
Nothing negative	40	39	34
Negative aspects	25	36	41
Migrants are laying claim to scarce benefits.	18	33	28
They have occupied our land.	1	5	7
They have occupied housing, of which there is a shortage.	5	17	10
They are ousting us from our jobs.	6	18	11
They cost the region too dearly.	4	4	4
They cost the enterprise or farm too dearly.	—	3	1
They are wealthier than we are, and yet they are asking for aid.	8	10	11
Certain traits of resettlers arouse antagonism.	9	12	25
They do not know how to work.	1	4	9
They do not want to work.	6	7	16
They behave badly.	5	2	6
Difficult to answer	35	24	25

— Not applicable (no members of group in the sample).

Russian rural residents are much more likely than townspeople to see something positive in the arrival of forced migrants (Table 7). However, this positive view is based on practical benefits ("They have recovered wasteland"; "They have built or repaired houses"; "They have improved the demographic situation thanks to an influx of young blood"), not on an assessment of the migrants' human qualities. Rural residents, more often than townspeople, say that migrants are close-knit and ready to help one another; they are less likely to mention their correct behavior, which they to a greater extent perceive as alien.

Migrants' Attitudes

Local residents' perception of a certain degree of self-isolation by resettlers has some foundation. There are two main causes for this separation: forced migrants' desire for solidarity with other migrants, and their negative attitude toward integration, based on unfavorable, stereotyped notions about the local population.

In surveys of relations between forced migrants and the local population based on cross-cultural psychological analysis, resettlers tend to evaluate themselves positively and local residents negatively (Gritsenko 1999: 54–61). Forced migrants have a need for psychological mechanisms aimed at consolidating the group and increasing the positive identity of its members. In part, migrants formed their high self-image while still in their former homes, as a result of their role as

Table 7. Positive Aspects of the Inflow of Forced Migrants according to Local Residents, by Type of Settlement

(percentage of local residents questioned)

Response	Large and medium-size cities	Small towns	Villages
Nothing positive	19	33	32
Positive aspects	35	37	46
They help improve the economic situation.	20	23	31
They have revitalized the enterprise or farm.	4	7	5
They have recovered wasteland.	10	11	22
They have built or repaired houses.	13	10	16
They are helping the local population.	1	2	—
They improve the demographic situation.	13	11	18
They are nice people who give the local population a good example to follow.	18	15	13
Difficult to answer	45	30	22

— Not available.

"culture bearers" (see Pilkington 1998). Such a self-appraisal was one reason for the conflictual character of their relations with the indigenous populations. To a certain extent, they have imported this conflict to Russia. It is the most pressing problem in small towns and rural areas, where highly urbanized resettlers perceive the local population as provincial and lacking in culture.

Resettlers were asked to name positive and negative qualities of local residents. Both types of quality were mentioned with approximately the same frequency (Figure 4). Among positive traits, kindness and a benevolent or "normal" attitude toward migrants stand out. On the negative side, 38 percent of the migrants mentioned drunkenness, and a total of 17 percent mentioned envy, gossiping, bad language, greed and thievery, and laziness.

Given such notions about the local population, it is not surprising that resettlers strive for a certain degree of isolation. They do not avoid contacts with the local population, but they prefer socializing within their own circle.

Alexander Alexeyev, a professor at Moscow State University, believes that integration, especially in villages, would be a disaster for resettlers.[7] In the course of the surveys, we witnessed situations in which migrants who had settled in the countryside adopted the social patterns that are widespread in Russia, so as not to remain outsiders. Quite often, the result was alcoholism and degradation. However, under the right conditions, integration could be a boon for the local population rather than a blight for resettlers.

Predominant among the positive qualities that the Russian population sees in forced migrants are unity, mutual assistance, and the willingness and ability to work. In both the 1997–98 and the Tver surveys, these characteristics were mentioned much more often than negative qualities. In other words, local residents particularly notice and positively assess qualities that resettlers acquired in other cultures and in response to the hardships they have had to undergo. It is thus possible to regard forced migrants as having great potential for leadership in restructuring the Russian population's value system. And a segment of the local sample does see such a role for this group.

---

[7] Alexeyev voiced this idea at the roundtable conference "Integration into Society of Migrants and Members of Ethnic Minorities," conducted in Moscow by the Ministry of Internal Affairs of the Russian Federation jointly with the Directorate of Social and Economic Affairs of the Council of Europe, November 18, 1998.

The share of answers to the effect that forced migrants set a good example for local residents remained constant and substantial (migrants are polite and respectful; they are hardworking; they spur people around them to be more active; and they are well educated and raise the cultural level). Answers to this effect were given by nearly one in six of those questioned. The results give rise to hopes for two-way traffic along the path of integration—even though such a path is long and difficult, especially given the present economic situation in Russia.

Policy toward Migrants

As a result of their unplanned emigration and a restrictive policy toward reception of migrants in large cities, forced migrants find themselves in small, closed communities, which are the ones most reluctant to receive newcomers. Migrants increase the population of depressed urban settlements, especially small towns, where their inflow exacerbates competition. Many are compelled to live in the countryside, where the sociocultural distance between the host population and the newcomers is the most pronounced. This settlement pattern increases the chances of mutual rejection. Those resettlers who are trying to adapt but who reject integration (which they see as jeopardizing their sociocultural identity) find themselves in a particularly difficult situation.

Although the three main circumstances that have a decisive impact on the adaptation of resettlers—employment, housing conditions, and relations with the local population—are more favorable in cities, current policy works to direct the forced inflows toward the countryside. Such an outcome worsens the conditions for forced migrants' adaptation and creates an obstacle for future migration. This is confirmed by the decreased migration inflows from the new independent states in the past three years and by the frequency of reverse migration, even though the situation in the places the migrants left has not improved and may have worsened.

Forced migrants in villages receive the most assistance in obtaining housing, and those in cities receive the least. The share of migrants who have received such assistance from the migration service is 1.8 times as large in villages as in cities, and the share of those who have been assisted by local authorities is 3 times as large. Enterprises and farms also have much greater opportunities to give assistance in small towns and villages.

One form of public assistance to resettlers established by law is interest-free long-term repayable loans for the construction and purchase of housing. Although the share of respondents who applied for the loans in different types of settlement was absolutely the same, only one fourth of the applicants living in cities and about one third of those living in small towns received loans, but more than half of those living in villages did (Table 8).

The receiving and assistance policies affect not only resettlers but also Russia as a whole. The country is losing substantial demographic and human resource potential under conditions of chronic depopulation and "brain drain," as well as considerable business potential that could have become an engine for economic reform. Simultaneously, social tensions and poverty are increasing.

It is obvious that the direction of forced migrants' resettlement must be changed. To do this, it is necessary to take such actions as removing the administrative obstacles that hamper freedom of movement and developing a system of municipal leasing of dwellings in large cities. First, however, it is necessary to change public opinion toward the inflow of resettlers and to understand the inflow not as a hindrance or as material for filling economic or demographic gaps but as a valuable resource that can help secure the future of Russia.

Table 8. Aid to Forced Migrants in the Form of Housing and Housing Loans, by Type of Settlement

(percentage of those questioned)

Form of assistance	Total	Large and medium-size cities	Small towns	Villages
Housing, total	28	20	29	35
Provided by migration service	11	9	9	16
Provided by an enterprise or a farm	11	8	14	12
Provided by local administration	4	2	3	6
Provided by a nongovernmental organization	2	1	3	1
Housing loans				
Applied for a loan	32	32	32	32
Received a loan	12	8	11	18

References

Arutyunian, Yu. V., and others, eds. 1992. *Russians: Ethnosociological Essays* [in Russian]. Moscow: Nauka.

Finansy i Statistika Publishers. 1990. The Educational Level of the USSR Population (According to the Findings of the 1989 All-Union Population Census) [in Russian]. Moscow.

Goskomstat (State Statistical Commission). 1998. *Chislennost i migratsiya naseleniya Rossiyskoy Federatsii v 1998 godu* [The numerical size and migration of the population of the Russian Federation in 1997]. Statistical Bulletin. Moscow.

Gritsenko, V. V. 1999. Russians among Russians: Problems of Adaptation in Russia of Forced Migrants and Refugees from the Former USSR Republics [in Russian]. Moscow: Institute of Ethnology and Anthropology, Russian Academy of Sciences.

Moscow Center of the Carnegie Endowment for International Peace. 1998a. *Politichesky almanakh Rossii* [in Russian]. Moscow.

———.1998b. *Poverty: Alternative Approaches to Definition and Measurement.* Moscow.

Pilkington, Hilary. 1998. Migration, Displacement and Identity in Post-Soviet Russia. London and New York: Routledge.

Vishnevsky, A. G., ed. 1998. *Population of Russia 1997: Fifth Annual Demographic Report* [in Russian]. Moscow: Institute for Economic Forecasting, Center for Demography and Human Ecology.

Vitkovskaya, Galina S. 1993. *Forced Migration: Problems and Prospects.* Moscow: Institute for Economic Forecasting of the Russian Academy of Sciences and RAND.

———.1998. *Resettlement of "Refugees" and "Forced Migrants" in the Russian Federation.* Geneva: Technical Cooperation Centre for Europe and Central Asia, International Organization for Migration.

The Rural Economy of the Tunka Valley in a Time of Transition and Crisis

Sergei A. Panarin

In 1992–95 a small research team composed mainly of scholars of the Institute of Oriental Studies, Russian Academy of Sciences (including the author) carried out a series of field studies in Tunka District, Cis-Baikal region. The aim of the studies was to shed light on how people in a rural locality—taken not as impersonal subjects of the reform policy but as independent actors—have adapted to the dramatic circumstances of transition in post-Soviet Russia. Each year during the period, the team spent at least one of the summer months living in the village of Tory and collecting field materials at the district and village levels. The fieldwork was followed by research in the archives and by data processing and interpretation.[1] In 1999, during the author's brief visit to Kyren, the district's administrative center, additional statistical data were collected at the district office. These data helped clarify the new demographic and economic tendencies that have emerged in the region since 1995.

Tunka and the Changes

The Tunka Valley lies to the west of Lake Baikal and stretches latitudinally between Irkutsk *oblast* (region) to the north and Mongolia to the south. The valley forms part of the Djida-Selenga agroclimatic region of the south Siberian mountain taiga zone and is notable for its relatively favorable conditions for agriculture (annual precipitation, 300–400 millimeters; frost-free period, 85–104 days) (BB RGS 1992: 67). From east to west, the valley is crossed by the Kultuk-Mondy highway, more than 250 kilometers long. The total area of Tunka district is almost 11,800 square kilometers. As of January 1, 1992, the population was 29,000, in 8,408 households; by January 1, 1999, it had fallen to 27,000.[2] Buryats and members of tiny aboriginal groups known as buryatized Soiyots and Evenks make up 52 percent of the population; the remaining 48 percent

---

[1] Before 1999, only the preliminary conclusions had been published (Panarin 1994, 1997). The 1997 paper, a short conceptual note, was later published in English translation (Panarin 1999).
[2] Statistical data from the Tunka local administration, Dossier "Labor."

consists of Russians and a small number of russified Ukrainians and Tatars.

A large part of the region's territory is covered with taiga untouched by logging. The region has ample recreation and tourist potential, as it is famous for its picturesque landscapes and healing thermal and other mineral springs. There is, however, only one functioning resort, Arshan, and one hydropathic establishment, Nilova Pustyn'. Thus, the Tunka region remains predominantly agrarian. It is only in the far west, in Samarta, that gold-bearing strata are being exploited. All of the settlements scattered in the valley, including Kyren, are rural.

Tory is the first big Tunka village one encounters when coming from the direction of the Irkutsk oblast border. It is situated on the Kultuk-Mondy highway, 55 kilometers from the southwest corner of Lake Baikal. According to the republic's statistical office, at the beginning of 1993 the village had 1,122 permanent residents in 349 officially registered households (SSC 1992: 3). In June 1993 the research team drew up a household plan of the village, at the same time carrying out a census of the local families, and found that there were 1,140 persons living permanently in Tory, in 305–310 households. Ninety-three percent of the families were of Buryat origin and 7 percent of Russian and mixed origin. The Russians in Tory exerted a negligible impact on village life.

Tory began to take shape as a unified village community in the 1930s. After 1955 strong consolidating factors appeared: a hard-surfaced road went through the village, and the central buildings of the Lenin collective farm, a rural council (sel'sovet) office, and a secondary school with boarding facilities were constructed on village territory. As a result, Tory came to be a center for the surrounding population. However, in the 1970s and 1980s the growth of the village slowed. By the 1990s a combination of a lower birthrate, a rising death rate, and an accelerating outmigration of young people caused the village population to stagnate at the 1993 level (1,128 as of January 1, 1999).[3]

Beginning in the 1930s, the valley and the village began to shed many aspects of the ancient Buryat and Russian traditions—the organic forms of economic activity and social organization, and long-established patterns of behavior and cultural creativity. This process intensified in the succeeding decades. From the beginning of the 1980s, however, signs of a reversal became ever more noticeable. The next decade

---

[3] Tunka local administration, Dossier "Labor."

proved to be a time of radical change in Tunka's status and the living conditions of its population.

In 1991 the whole territory of the region was enclosed within the Tunka National Park, which has autonomy from the district government.[4] Some restrictions were introduced to limit economic activities that could damage the environment. At the same time, under the political and economic transition, the village communities in Tunka were, in effect, made more or less independent social actors. They immediately faced a problem of survival because the state did not so much liberate them as cast them to the winds of fate. Tunka and Tory had to adapt, and they responded by turning to pre-Soviet traditions. The mainstay of the adaptation process has been the small-scale peasant economy that grew out of the collective farmers' household plots, *lichnoye podsobnoye khozyaistvo* (LPKh).

The state of Tunka's economy generally mirrors the condition of Buryat animal husbandry. Therefore, while the major findings are aggregated at the district level, some of the analysis is based on an in-depth study of Tory, where the Buryats are dominant in every aspect of economic activity and social life. That accounts for the emphasis on the Buryat component of the local population in this paper.

Two Historical Forms of Economic Organization in Tunka

Even if it is a pillar of adaptation, the LPKh is not the sole framework for productive activity. In fact, the process of adaptation relies on two forms of economic organization, differing in their origin and technological characteristics: the traditional and the modern models. To demonstrate the nature of the two models and how they coexist, I first present them as ideal types and then discuss deviations from the ideal resulting from their interaction and the effects of the social environment.

The Traditional Pattern

The Buryats have a centuries-old tradition of seminomadic cattle breeding. Before 1917 the Tunka Buryats had already become significantly more sedentary than originally, adopting some crops cultivated by their Russian neighbors such as rye, oats, wheat, and barley. They also practiced hunting, fishing, and logging (on a small scale and only for their own needs), stored cedar nuts, and gathered herbs and berries.

---

[4] Resolution of the Council of Ministries of the RSFSR, May 27, 1991, no. 282. For its contents, see *Tunka: istoriya i sovremennost'* [Tunka: Past and present]: 303 (Ulan-Ude: Buryaad unen).

By and large, the majority of the Tunka inhabitants retained specific features of the Buryat ethnic culture. They bred local varieties of animals that were low in productivity but resistant to local diseases and climatic vicissitudes. Their herds were mixed, consisting generally of cattle, horses, and sheep: in 1928 the ratio between the first two and the third species was 4 to 1.[5] Their animal husbandry techniques were primarily adapted to the local ecology. The main production units were individual holdings (*arat*), clustered in small, dispersed settlements (*ulus*) with 30 to 35 households per settlement.[6] Their social security was ensured by a network of inherited affiliations perceived in terms of common origin. Linguistically and culturally, they were part of the Mongolian world, and they shared both Buddhist and shamanist beliefs, although some of them were baptized.

East of Lake Baikal, the Buryat herds fed at pasture nearly all year round. In Tunka, however, a heavy snowfall was not a rare visitor. Indoor winter maintenance of the animals included limited outlays of fodder that covered, at best, not more than one-third of the amount required.[7] The necessity of pasturing determined the low density of the population. Every ulus had various economically significant areas that provided a fodder base. There were homestead plots and open meadows for mowing, and neighboring land and distant areas for pasturing. The latter were sometimes up to 250 kilometers away from the ulus, but usually they were much closer, in the foothills of the Sayan. All this permitted land tenure to vary according to the seasons and allowed each plot to be used for its most valuable product. At the same time, it helped prevent overuse of the limited areas. Manuring of the meadowlands by the grazing herds was widely practiced; the fertilized lands (*utug*) were subsequently cultivated.

The low yields and weight gain of cattle (compared with those achieved by modern farmers) were offset by high livestock numbers. A numerous multispecies herd had a triple function. It provided a hedge against natural calamities; it allowed farmers to use the natural resources of the valley in an effective way, avoiding depletion; and it defined personal status, thus taking on a social function of paramount im-

---

[5] Calculated from BB RGS (1992): 68, 75.

[6] Calculated from NARB (National Archives of Republic Buryatia), Deposit R-362, Inventory 1, Dossier 8, Sheet 28; Dep. R-475, Inv. 1, D. 368, Sh. 72.

[7] For the Trans-Baikal territory. the ratio was 15–20 percent. See Manzanova and Tulokhonov (1994): 103. Due to heavier snowfalls in Tunka, the local arats had to store more hay.

portance. All this meant that having a big multispecies herd was the predominant goal for the Buryats.

Agriculture was more significant for the Tunka Russians than for the Buryats, but otherwise the domestic economy systems of the two ethnic groups were similar. On the eve of collectivization this system sustained about 45,000–46,000 cattle, 15,000–16,000 horses, and 15,000 sheep (BB RGS 1992: 68, 75), which met the basic food requirements of the population of the valley. On April 25, 1928, the population totaled 24,600, slightly less than in 1992.[8]

The Modern Form

The situation depicted above is generalized and "ideal." It had its reverse side: a low level of literacy, poor sanitation among Buryat families, lack of modern public health services, and, as a consequence, high maternal and child death rates.

In the course of socialist modernization, the people of Tunka were aggregated in large production collectives—*kolkhozy* and *sovkhozy* (collective farms and state farms)—and in compact settlements that resembled Russian villages. With collectivization, the sown areas began to expand, from 10,000–11,000 hectares in 1928 to 19,000 hectares in 1953. In 1974 the total area of arable land in the valley was more than 41,000 hectares. Over 20 years the fallow share had decreased from 29 to 11 percent, leading to land degradation. The expansion of grain agriculture was accompanied by changes in the crop pattern: from the mid-1950s wheat occupied 60–70 percent of the total sown area, whereas its share before collectivization did not exceed 15 percent.[9]

During the collective period the number of horses drastically decreased, from 16,000 in 1928 to 3.700 in 1976. The number of sheep grew to 26,000 (BB RAS 1992: 69, 74–75), and the ratio of cattle and horses to sheep sank to 1:1. The cattle were improved as dual producers through introduction of Simmental blood, and the Buryat coarse-wool sheep were partly replaced by fine-fleeced breeds.

The collective cattle were partly transferred to stalled maintenance. The share of open pasture forage, which once provided two thirds to three fourths of the annual ration, declined to 15–25 percent, and the share of stockpiled, processed feed increased to 75–85 percent.[10]

---

[8] NARB, Dep. R-475, Inv. 1, D. 368, Sh. 72.
[9] NARB, Dep. R-475, Inv. 1, D. 480, Sh. 25; BB RAS (1992): 74–75.
[10] Calculated from IZC (1978): 22.

The state-supported system of social insurance was extended into the rural setting, along with public health and education, official secular culture, and the standard Soviet institutions of local power and Communist party control. But the new "ideal type" failed to supersede the traditional one altogether. Instead, the ancient pattern of Buryat life was restructured in a dualistic way, as grain production supplemented and in some ways competed with animal husbandry.

Grain was raised by extensive methods, although the main production operations were mechanized. In animal husbandry, intensive development was chosen as the principal model. The concept, however, was not fully carried out: highly productive cattle breeds were not given the needed care, as a consequence of the extremely low level of mechanization in this branch of the economy. In 1976, at the Lenin collective farm, only 28 percent of milking and manure removal was mechanized (IZC 1978: 32). A mixed or interim model of keeping animals partly in stalls and partly at pasture was created, and this hybrid proved both capital- and labor-intensive.

Because mechanized agriculture was cherished by the state while animal husbandry was neglected, there appeared a kind of social division between the two branches: the best-qualified, physically strongest, and hardest-working people were concentrated in the grain-producing sector, while those with an exactly opposite set of qualities gravitated to animal husbandry (see Humphrey 1983: 180–81).

Collective production, because of its inadequacies, failed to displace individual production on personal plots. Labor in the framework of the collective production system did not provide for more than 50 percent of the total family income of the typical collective farmer; personal plots accounted for the rest (IZC 1978: 73). The household production process straddled awkwardly over the two sectors of the rural economy. Both types of holdings competed for fodder and labor. In this race, collective production could never gain a definitive victory over individual production; such a victory could be guaranteed only by compulsion.

In 1983 holdings in the valley were exempted from having to make grain deliveries to the state. The fallow area could be doubled, and wheat could be replaced with a vetch-and-oats mixture. A barter payment system was introduced in field-crop cultivation. Thanks to these measures, plant cultivation was put to work for animal husbandry, and cultivators consequently had a vital stake in the results of their labor. The changes of the 1980s softened the modern form of economic organization, and the production process reached its optimum.

On the Lenin collective farm the productivity of grain crops improved from 15–17 metric quintals per hectare at the beginning of the 1980s to 21–23 metric quintals at the beginning of the 1990s. (A quintal is about 100 kilograms.) Other evidence of optimization is the formation of an entire stratum of families (17.3 percent of the total number) whose money income, gained on a collective farm or in the state employment system, allowed them to forgo keeping livestock.[11]

Such prosperity, however, rested on unsteady foundations. First, the modern form was not self-sufficient. There was a continual need for supplies of centrally provided technical devices, fuel and lubricants, mineral fertilizers, insecticides, elite semen, and so on, and these needs were growing at a higher rate than output and labor productivity. Second, the technology that gained a firm hold in Tunka contributed to degradation of the soil. At the end of 1990, 63 percent of the total arable land of the region and 6–8 percent of the pastures and meadows were suffering from water and wind erosion (BB RGS 1992: 74). Third, the crop and livestock branches continued to develop unevenly. Compared with the relatively tangible well-being of arable farming, animal husbandry was stagnating. From 1976 to 1991 on the Lenin collective farm, average milk yield per year per forage cow remained practically the same: 1,615–1,625 liters.[12]

Modernization made possible the social progress of the Tunka Buryats, but it did not afford economic solvency. At the beginning of the 1990s the population of the Tunka Valley was almost the same as on the eve of collectivization, and industrialization and urbanization had bypassed Tunka. Thus, with the same population numbers and the same settlement and employment patterns, total livestock holdings in the individual and collective sectors in 1991 were, for cattle, 83 percent of the 1928 number; for horses, 38 percent; and for sheep, 41 percent.[13] Had 1986 been taken as the reference year, the figure for sheep would have been 150 percent, but in the late 1980s and early 1990s great numbers of the sheep in Tunka kolkhozy and sovkhozy were slaughtered. This was done to comply with the recommendations of the republic's agricultural authorities, which sought to reduce overpasturing in the region. Maintenance of the ecological balance in the Tunka National Park was also a consideration. Even bearing that in mind, the fact remains that at best, under the modern form, the total number of animals was no greater than

---

[11] Field data collected in 1993.
[12] IZC (1978): 29; field data collected in 1995.
[13] Calculated from BB RGS (1992): 68, 69, 75.

before collectivization, while the herd structure had deteriorated and the resources needed for livestock raising had been depleted.

Principal Features of the Current Crisis

Since 1991 the state has largely withdrawn support from agriculture. The sovkhozy and kolkhozy of Tunka have been transformed into loosely structured associations of peasant farms (APF) that have inherited some of the managerial and distributive functions of their predecessors but have failed to replicate their size and efficiency. Only the Lenin collective farm and the Turansky state farm did not break up.

The few private farmers who appeared in the valley at the dawn of the reform are faring worst of all. Some tens of them are specializing either in arable agriculture or in rearing animals that are exotic for the Buryats (geese, rabbits, and polar foxes). These farmers' share of the number of cattle does not exceed 2 or 3 percent. By contrast, according to official data, between January 1, 1991, and June 1, 1999, the share of peasant households (LPKhs) in the total number of the main type of livestock (cattle) increased from 35 to 83 percent.[14] Household farms were exempted from fiscal and administrative restrictions on livestock numbers and were granted ownership rights to their plots. It seemed that a significant part of them would gradually evolve into true commodity producers of animal husbandry products.

The field data collected in 1992–95 and the statistics available for 1996–99 do not support such an optimistic prognosis. The pronounced change in the ratio between private and collective herds was attributable more to the decrease of collective herds than to the growth of private herds. The kolkhoz and sovkhoz cattle herd declined dramatically; between 1991 and 1995 it fell from 24,926 to 11,868, and it had dropped to 2,947 by June 1, 1999.[15] In 1991 in the collective sector there were more than 3,300 sheep and goats and about 4,500 pigs. By June 1999 the numbers of livestock in individual peasant households were, for cattle, 17,716 (71 percent of the number of kolkhoz and sovkhoz cattle in 1991); for sheep, 1,810 (55 percent); and for pigs, 1,759 (less than 40 percent).[16] Thus, the growth of the peasant household herds did not make up for a swift decrease in collective livestock. As the statistics reveal, the total number of animals in individual peasant households and the average annual rate of growth of livestock look rather modest (see

---

[14] Calculated from statistical data from the Tunka local administration, Dossier "LPKh."
[15] Tunka local administration, Dossier "LPKh."
[16] Tunka local administration, Dossier "LPKh."

Tables 1 and 2). During 1991–97 the average annual growth rate for total number of livestock was only 3 percent.

Table 1. Numbers of Livestock Held by Individual Peasant Households, 1991–97

	1991	1992	1993	1994	1995	1996	1997
Cattle	13,417	14,446	15,011	14,202	15,129	16,368	16,296
Cows	5,551	6,146	6,418	6,533	6,819	7,164	7,182
Horses	318	476	640	703	833	1,037	1,198
Mares	134	188	256	346	386	358	412
Sheep	2,770	3,220	3,057	3,162	3,092	2,692	2,206[a]
Pigs	3,331	4,054	2,516	2,061	2,248	2,481	2,122
Poultry	4,400	4,900	6,125	7,106	8,964	9,956	8,519
Total[b]	12,212	13,519	13,552	13,064	13,920	14,950	14,800

Note: Data are as of January 1 for each year.
a. Includes a small, unspecified number of goats.
b. One cow =1; other cattle = 0.6; 1 mare = 1; other horses = 0.6; 1 sheep = 0.16; 1 pig = 0.35; 1 fowl = 0.02.
Source: Statistical data from the Tunka local administration.

Table 2. Growth Rate Index for Individual Peasant Households, 1991–97

(1991 = 100)

	1991	1992	1993	1994	1995	1996	1997
Cattle	100	107	112	106	113	122	121
Cows	100	111	116	118	123	129	129
Horses	100	150	201	221	262	326	377
Mares	100	140	191	258	288	267	307
Sheep	100	116	110	114	112	97	80
Pigs	100	122	76	62	67	74	64
Poultry	100	111	139	161	204	226	194
Total[a]	100	111	111	107	114	122	121

a. One cow =1; other cattle = 0.6; 1 mare = 1; other horses = 0.6; 1 sheep = 0.16; 1 pig = 0.35; 1 fowl = 0.02.
Source: Table 1.

The official census of privately held animals is conducted at the beginning of the year. Generally, it is perceived as a fiscal measure, and all householders underreport the number of animals they actually possess. At the end of spring and the beginning of summer, however, when animals should have veterinary checks, the owners show a lively interest in having all their animals seen, and the administration of the region gets a chance to correct the winter data. As a result, the total number of livestock in individual peasant households jumps by more than 30 percent. Unfortunately, we have succeeded in obtaining corrected data only for 1994 and 1999, and it is impossible to construct a statistical series. The underreported quantity, however, is more or less constant, and in practice it does not affect calculation of the growth rates. The latter were uneven: the number of cattle fluctuated; the numbers of sheep and pigs in individual peasant households have been decreasing, on the whole, since 1992; and the number of horses declined sharply after 1997.[17]

In order to explore the development tendencies and perspectives of the LPKhs, in 1995 the team conducted in-depth interviews in Tory with five experts (the collective farm chairman, head veterinary surgeon, veterinary surgeon for private animals, chief engineer, and chief economist), as well as 20 owners of individual households, selected by the experts as "well-off managers." For the household sample, we developed a loosely structured questionnaire to reveal households' main economic characteristics: total number of persons and number of workers; number of animals by species and category; form of landholding and use of land; technical equipment available; fodder stockpiling and production output for the previous year; and market sales of the products. The analysis of the resultant data confirms that the actual number of livestock in an individual peasant household is higher than that given by the statistics. In the 20 LPKhs there were 414 animals of different species, of which 41.5 percent were cows; 4.1 percent, horses; 21.5 percent, sheep; and 32.9 percent, pigs.[18] The ratio of the total number of cows and horses to the number of sheep was 2:1.

Except for one 30-year-old owner, all respondents, as well as the experts, agreed that it would be impossible to manage an individual holding without backing from a collective farm. A larger number—five owners—thought that they could make their way without help from family and friends.[19] Thus, an individual holding was found to be eco-

---

[17] The average annual growth rate would be cut in half if 1992 rather then 1991 were used as a benchmark. The abrupt change in the number of LPKh livestock that took place between January 1, 1991, and January 1, 1992, is attributable not only to real growth but also to a decrease in underreporting caused by the general liberalization of the economy after 1991.

[18] Questionnaire for 20 households.

[19] Questionnaire for 20 households.

nomically less dependent on the traditional form of social support than on help rendered by the modern form of organization. One explanation might be that kinship and neighbor relations are generally used for labor exchange and, to some extent, for exchange of ready-made goods, while the collective farm provides a considerable part of the production inputs for an individual peasant household. In fact, all householders receive from the collective farm (as part of their wages) fodder grain, mixed fodder, and straw.

The largest household landholdings are 2 hectares, but most landholdings are 1.5–2 times smaller. Such a plot allows a householder to keep three or four head of cattle and not more than two milk cows, whereas not less than three cows are needed to provide for the simple reproduction of a household. (Reproduction, in this context, means that that household is able to meet its basic needs as a production and consumption unit.) Ten of the households surveyed have only the minimum number, but five households have as many as four or five cows.[20]

If all the land belonging to the Lenin collective farm had been equally divided among its members, each would have had about 7 hectares of land of various qualities.[21] In that case only the smallest households would have had adequate supplies of natural fodder. But in reality, even their reproduction process would have been hampered because the cattle species that gained a firm hold in Tunka in the 1960s and 1970s need enriched feed to maintain productivity. Without the supplemental feed provided by the collective, it would be necessary to increase livestock numbers, and there would again not be enough natural fodder.

Thus, as shown by the 1992–95 data, the collective farm resources continue to be converted into production inputs for individual peasant households. But what can LPKhs offer the collective farm in return? Their labor in no way compensates for the equipment, mineral fertilizers, insecticides, and so forth formerly provided by the state. As a result, even grain agriculture is shrinking year by year and at present can scarcely support the LPKhs effectively with fodder, as it did five or six years ago. The average productivity of grain crops in Tunka dropped from 17 metric quintals per hectare in 1986–90 to 5 quintals in 1996–98, while the sown area decreased to one third of the original level.[22]

Another negative factor that affects the survival of the LPKhs is their limited access to markets for their products. The overwhelming

---

[20] Questionnaire for 20 households.
[21] Land data from the Tory *sel'sovet*.
[22] BB RGS (1992): 68, 76; statistical data from the Tunka local administration.

majority of LPKhs are exclusively consumer units, as indicated by their motivations, production structure, and volume of production. In our sample of 20 households, only 2 could be categorized as evolving towards a commodity economy, and only 5 householders hoped to enlarge their production.[23] However, even the smallest LPKhs, working exclusively for family consumption, are forced to go to market every now and then to make up for an acute shortage of cash in hand.

Conclusions

My initial hypothesis was that in the economic sphere, the elementary process of household survival would prevail both in the individual village and in the valley as a whole, encouraging the rebirth of traditional social relations and giving rise to intracommunity factional strife over resources and power. A limited growth of the private farm sector was presupposed. Taken together, all those changes would have marked a path of adaptation of the Tunka peasants to the constraints of economic and social transition in Russia.

My final conclusions differ to some extent. It is now clear that, except in some minor cases, private farmers have failed to flourish or even to come into existence. The peasant economy does show high resilience, as demonstrated by the growth of livestock numbers and the involvement of many households in food markets in the industrial towns of the Irkutsk region. However, the overwhelming majority of households linger in the folds of the subsistence economy, their sales of meat and milk being aimed at meeting family consumption needs. Even households with large incomes can maintain their relative prosperity only insofar as they have privileged access to the productive assets of a kolkhoz or an APF. Those excluded often cannot achieve consumption above the poverty line.

As a consequence, Tunka rural communities are now becoming an arena of factional struggle, the most coveted prizes being fodder and forage, either produced by or formally controlled by a kolkhoz, sovkhoz, or APF. The competition for resources and the search for social support encourage the reemergence of quasi-traditional social relations and shamanist rites. The former help with survival; the latter promote a milieu for consolation and assistance. The newborn or reborn economic forms depend on the persistence of the kolkhoz-sovkhoz system or, more exactly, of its remnants, which are doomed to a gloomy fate as a consequence of noninvestment by the state and the sharp decline of la-

---

[23] Questionnaire for 20 households.

bor inputs from peasants. If these farms definitively collapse, many peasant households will perish.

From the political angle, the results of our study expose an acute need for careful rethinking of the role of collective enterprises. For the time being, it is necessary not to interfere with the spontaneously attained delicate balance of interests inherent in current interactions between such enterprises and peasant households. A danger, however, is that if the existing pattern of nonrecompensed appropriation of collective resources persists, Tunka is likely to experience a new wave of economic recession.

References

BB RAS (Buryat Branch of the Russian Geographical Society). 1992. *Tunkinskii natsional'nyi park (priroda—khozyaistvo—naseleniye)* [The National Park of Tunka: Nature—economy—population]. [Ulan-Ude]. Manuscript.

Fadeyeva, A. 1997. "Tunka–Oka: dalyokiye, no ne broshennyye [Tunka and Oka: Remote but not abandoned]. *Buryatia* (Ulan-Ude), April 8.

Humphrey, Caroline. 1983. *Karl Marx Collective: Economy, Society and Religion in a Siberian Collective Farm.* Cambridge, U.K.: Cambridge University Press.

IZC (Irkutsk Zonal Center for Organization of Labor and Production in Agriculture). 1978. *Plan sotsial'nogo i ekonomicheskogo razvitiya kolkhoza im. Lenina Tunkinskogo aimaka Buryatskoi ASSR na period 1976–1990 gg* [The plan of social and economic development for the Lenin Collective of the Tunka District, Buryat ASSR, 1976–1990]. Irkutsk. Manuscript.

Khazanov, Anatoly, Vitaly Naumkin, Kenneth Shapiro, and David Thomas, eds. 1999. *Materialy mezhdunarodnoi nauchnoi konferentsiyi "Sovremennoye sostoyaniye skotovodstva i zhivotnovodstva v Kazakhstane i perspektivy ikh razvitiya,"* Almaty, 12–13 yanvarya 1999 g. Moscow: University of Wisconsin (Madison), Russian Center for Strategic Research and International Studies.

Manzanova, G. V., and A. K. Tulokhonov. 1994. "Traditsiyi i novatsiyi v razvitiyi sel'skogo chozyaistva Zabaikal'ya" [Agricultural development in the Trans-Baikal territory: Traditions and innovations]. *Vostok/Oriens* (Moscow), no. 1.

Panarin, S. A. 1994. "Tunka na pereput'ye: nabrosok k sovremennoi sotsioyestestvennoi istoriyi Pribaikal'ya" [Tunka at the crossroads: Notes on the contemporary social-cum-environmental history of the Cis-Baikal region]. In *Materialy vtoroi nauchnoi konferentsiyi "Chelovek i priroda—problemy sotsioyestestvennoi istoriyi"* [Proceedings of the 2nd Conference on "Man and Nature—Problems of social-cum-environmental history"]: 106–21. Moscow: Moskovskii litsei,

———.1997. "Buryatskoye selo Tory v 1990-kh gg: sotsial'naya i kul'turnaya readaptatsiya maloi poselencheskoi obshchnosti" [The Buryat village of Tory in the 1990s: Social and cultural readaptation in a small village community]. In *Rossiya i Vostok: traditsionnaya kul'tura, etnoku'turnyye i etnosotsial'nyye protsessy* [Russia and the East: Traditional cultures, ethno-cultural and ethno-social developments]: 149–51. Omsk State University.

———.1999. "The Buryat Village of Tory in the 1990s: Social and Cultural Re-adaptation in a Small Village Community." *Inner Asia* (Cambridge, U.K.) 1(1): 107–10.

SSC (State Statistical Committee of the Buryat Republic). 1992. *Sel'skoye naseleniye Respubliki Buryatiya na 1 yanvarya 1993* [Rural population of Buryatia as of January 1, 1993]. Ulan-Ude.

This chapter is based on the paper "Khozyastvo traditsionno skotovodcheskogo raiona v period krizisa: opyt buraytskogo sela Tory " [Farming in a traditionally cattle-breeding area in the period of crisis: The experience of the Buryat village of Tory], originally presented at the conference on "The Present State of Cattle-Breeding and Animal Husbandry in Kazakhstan and the Prospects of Their Development," Almaty, January 12–13, 1999, and published in Khazanov and others (1999): 198–219. I would like to thank the editors for kindly permitting me to use the original version. Because new statistical data for the period 1996–99 were included; the paper was restructured and reformulated to a great extent.

Studying the Political Economy of the Rural Transition

L. Alexander Norsworthy and Alexander Feshenko

In many ways, looking for continuity in Russian events is more important than envisioning dramatic departures from established patterns of institutional behavior as the result of an election. Typically, a change in regime involves more continuity than radical transformation (Suzuki Norsworthy and Brough 1992, Suzuki, Norsworthy and Gleason 1993). In fact, the Imperial and Soviet experiences proved that reforms were cyclical.(Norsworthy 1989) As the papers in this volume illustrate, rural reform in Russia has been obstructed by structural, policy, and behavioral forces---vestiges of both the Soviet system and land tenure arrangements in imperial Russia. Plots and large farms (jont stock companies or collectives) emain the dominant production units. The private, "independent" farms have not emerged as a significant force during the rural transition. In short, there is not yet a powerful organized group of production units (in terms of arable land controlled or agricultural output) occupying the gap between the successors to the *kolkhozy/sovkhozy* and the households. The former operate in many of the same ways they did prior to the reforms, and the advantages of the new land ownership regime have not benefitted the plots held privately over those held collectively.

These obstacles to making private property and the market the dominant factors in the rural sector need to be studied within a framework that reflects the realities of the current situation. This compendium has been organized using a framework which is based on studying structures, policy outcomes and adaptive responses.

Institutions

On the institutional front, the events of late 1999 illustrate some important realities. A coalition of interests that formed within the last two months before the Duma elections in December 1999 was responsible for the elevation of Vladimir Putin. Yeltsin's resignation suggests that the pervasiveness of corruption had called into question the building of long-term relationships of ranking agriculture officials in Russia with the West and the international development community.

The immense popular support for the acting president signals the continuing power of hard-line regional politicians whose support he has actively pursued. The Communist Party of the Russian Federation (CPRF) received the largest number of seats in the Duma elections, although it does not have a majority. Reform can proceed, but only with its blessing. The Agrarian Party—a coalition of interests, rather than a fully institutionalized political party—won no seats in the Duma—it was not necessary the hastily assembled electoral coalition was already representing their interests

Many Washington pundits had projected a breakup of the Russian Federation at the time of the adoption of the new Russian Constitution. The war in Chechnya illustrates some important points regarding Russian federalism. The federal government's responses to events in the autonomous republic have demonstrated the primacy of national interests and that Russia will not abide foreign interference in its internal affairs—at least not when world oil prices generate sufficient revenue for the regime to take a go-it-alone approach.

Policy Outcomes

Category B enterprises—in this case, individual plots, whether household plots or garden plots allocated by a joint-stock company, a collective farm, or the municipality—account for a disproportionate share of agricultural output, given the minuscule amount of land they represent.

Uzun suggests that the logical extension of the role of the collective farms and their successor entities should be to act as service and input suppliers for "independent" farms. If the strict definition of Category C is used (larger than a houshold plot, privately held) there is not enough demand for the services from such a small segment of producers. Clearly Uzun is referring both to the private farms as well as to the household plots. Also, such a scenario assumes that managers will not seek to use their access to political leaders and their control over the distribution of inputs and subsidies to Category B and C entities to perpetuate their advantages in political access and subsidies at the expense of the latter.

The answer lies in mobilizing Category B and C entities in the rural sector and using nongovernmental organizations (NGOs) and the sustainable self-financing mechanisms at the NGOs' disposal to increase awareness of property rights, improve the transparency of land transactions, and move informal activities progressively out of the gray economy and into the formal sector through advantageous tax treatment.

This would be political action for the smallholders, by the smallholders. It is unclear to what extent multilateral development banks and bilateral or multilateral aid agencies can facilitate these transformations without grassroots impetus from the rural communities themselves.

Category A enterprises may benefit from the new openness; they may be able to maintain the status quo through strenuous effort; or their managers may simply have to watch the erosion of their assets. The economic behavior of the larger Category A entities depends on the labor pool provided by households. As Kalugina's analysis indicates, the enterprise's commitment to its labor force is a crucial element in the firm's prospects. Those enterprises that seek to retain their workers and maintain the social infrastructure by providing the same level of services as before have the best record of economic accomplishment. Where the management has been inclined to strip the joint-stock company of assets, the labor force has suffered great hardships.

The agricultural joint-stock companies and the remaining state and collective farms have severe corporate governance problems. Accountability to shareholders is not uniformly the motivating factor for managerial performance, nor is it a criterion for managerial compensation. The distribution of land and capital shares is not transparent, and there is insufficient regulation of the enterprises to ensure that shareholders are fully aware of and able to exercise their rights of voice and exit.

As the migration studies by Vitkovskaya and Panarin illustrate, the costs of exit from a community are high. The demographics of the rural population are unfavorable, with elderly people who are dependent on pensions and other invisible subsidies for their sustenance constituting a major population group. The social fabric of rural communities makes it possible, however, for households to sustain themselves through the barter economy and other informal transactions. New arrivals in rural communities do not so much face discrimination as suffer from a lack of connections. Access to inputs and resources is the key to economic survival and a newcomer may not enjoy the same level of access as lifelong residents. In this sense, little has changed from the Communist period, when access was the currency (see Voslensky 1984).

Adaptive Responses

The survey data used by Bogdanovsky, Serova, Artemova, and other contributors to this volume indicate a general malaise among opinion leaders, decisionmakers, institutions, households, and individuals. To approach these issues, the level of analysis needs to be clearly defined.

Things are not expected to get better for the employees of Category A enterprises. Many of them, however, may well have their own plots, putting them in Category B or C. Category B or C respondents are more positive than are workers who have no prospects of moving beyond subsistence farming, as is indicated by data from VTsIOM, the All-Russia Center for Public Opinion.

The most destabilizing developments that need to be studied within an integrated analytical framework are the migration of residents out of rural areas and immigration by residents of the new independent states and by illegal aliens. The influx is straining already decaying social infrastructure. Eventually, groups representing the interests of these immigrants may articulate political interests and demands on their behalf.

Toward a Household-Based Rural Political Economy

To draw on these lessons from Russian scholars, indicators need to be established to measure inputs, outputs, and the impact of policies and investments. The analysis needs to take into account three institutional levels—national, oblast, and municipality—and the division of authority over resource allocation at each level . The following questions are suggested topics to be addressed in subsequent research.

Institutions. Is it possible to measure the success of a reform package without accounting for political variables? How can a reform package be proposed and interventions funded if the necessary institutions are not already in place? What are the appropriate institutions for a given set of reforms in a particular milieu? How can the readiness of these institutions be systematically assessed?

Policy outcomes. Can the success or failure of a reform package be judged solely by whether laws are passed, or should there be an effort to understand the vehicles for implementation, and the process of implementation, at the regional, provincial, district, and local levels? Is there a commonly agreed-on measure of poverty that is robust and is based on sound monitoring mechanisms?

Adaptive responses. How can public opposition to reforms be understood from a historical and sociocultural perspective? What remedies can be found to overcome these obstacles through reference to sociocultural institutions and use of grassroots organizations? What are the boundaries between public behavior and the private behavior of households and social networks? Are these boundaries changing, and, if so, are there measurable economic consequences?

Sector-specific approaches deal adequately with the "policy outcome" axis of analysis. The World Bank observes neutrality in political matters, but it has to be remembered that a wide range of activity may be considered "political" in Russia. For example, enabling participatory approaches to the design and implementation of projects is perceived as a political act. A community meeting to discuss postreform shortfalls in the provision of health care and child care services by a former collective farm may be politically acceptable in one oblast and not in another. The results of such a meeting may therefore not represent accurately the views of the stakeholders. Participants might be more forthcoming in personal interviews or focus groups, or they might withhold information or slant their responses because of perceived threats of retaliation, depending on their reading of the political context. Observations of behavior are at least as important as statements in public forums.

Methodological Issues

To overcome these analytical problems, theory building is necessary to elaborate a paradigm that aids understanding of political, economic, and social issues and that grounds policy recommendations on an integrated understanding of these three dimensions. Such a model requires the involvement of members of the particular society's intelligentsia who are familiar with common linguistic referents and can decode responses from interviewees and survey participants.

In order to more fully understand the behavior of Russian farmers, collection of detailed data on the informal sector in rural communities is needed. The types of transactions associated with the complex interrelations discussed by Amelina need to be categorized and studied in terms of their impact on firm and household behaviors.

The boundary between households and enterprises is perhaps the most difficult to assess and will vary according to survival strategy. as Kalugina suggests. Amelina points to the fluidity of the boundaries between firms. Similarly, the activities of a household member as a joint-stock company employee and as a private producer on a household plot obscure the line between entities in Categories A, B, and C. What the contributors to this volume make clear is that Category B and C entities need the large Category A farms to survive. The output of Category B and C farming suggests that this group of entities fills an important gap, particularly in vegetables, fruit, livestock and dairy products.

Households emerge from a review of the Russian papers as the central units of the social networks that hold rural communities and their economies together. The formal and informal transfers between households, their provision of labor to different rural enteprises or entities, the transfers from the government or enterprises to the household, whether transfer payments, wages or inputs for household farming are the essential transactions determining enterprise, group and individual behavior. (Bamberger, Kaufmann and Velez 2000)

Figure 1. Interconnectivity of Rural Entities

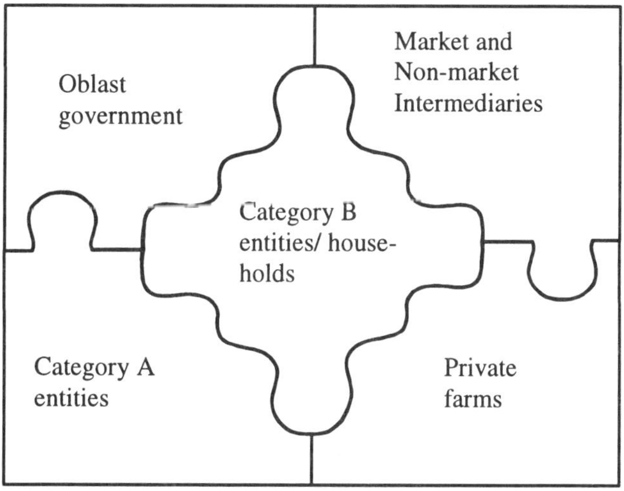

Figure 2. Household Inputs into Different Enterprises and Entities

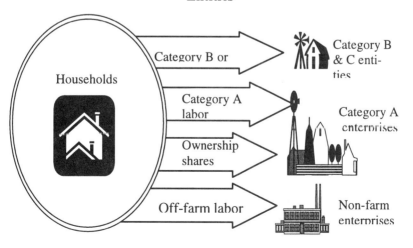

Conclusion

Both qualitative and quantitative research should be used to develop a holistic view of the survival strategies of enterprises and households. Qualitative assessments are needed to determine whether laws and rules are transparent. Failure to exercise rights can be remedied, and loss of property because of inadequate knowledge of those rights or because of authorities' unwillingness to enforce them can be avoided. However, Russia is not a society founded on the principle of private property like the United Kingdom or the United States. It is not sufficient in Russia to pass a law and expect it to be implemented or to grant members of the population a certain set of rights and expect them to exercise those rights. The Nizhny Novgorod pilot demonstrated that a complete set of public communication and education strategies must be implemented if households are to develop a sense of empowerment and if rural reform is to be implemented effectively.

Bibliography

Arutjunjan, Ju. V. *La structure sociale de la population rurale de l'U.R.S.S.* 1979. Annales de l'Institut national agronomique, Paris-Grignon: Sciences sociales; no 1. Paris: Institut de sciences mathématiques et économiques appliquées.

Bamberger, Michael, Daniel Kaufmann and Eduardo Velez.. "Interhousehold Transfers: Using research to Inform Policy." *PREM notes*, World Bank Group, 2000.

Blum, Jerome. 1961. *Lord and Peasant in Russia, from the Ninth to the Nineteenth Century.* Princeton, N.J.: Princeton University Press.

Burds, Jeffrey. 1998. *Peasant Dreams and Market Politics : Labor Migration and the Russian Village,1861–1905.* Pittsburgh, Pa.: University of Pittsburgh Press.

Kotsonis, Yanni. 1999. *Making Peasants Backward: Agricultural Cooperatives and the Agrarian Question in Russia, 1861–1914.* Houndmills, U.K.: Macmillan; New York: St. Martin's Press.

Norsworthy, Leonid A. 1989 "The Political Economy of Economic Reform and Technology in the German Democratic Republic: Structures and Policy Outcomes" Washington, D.C.: The American University/UMI.

Orlovsky, Daniel, editor.1995 *Beyond Soviet Studies* [Washington, D.C.] : The Woodrow Wilson Center Press ; [Baltimore, Md.] : Distributed by John Hopkins University Press

Pallot, Judith. 1999. *Land Reform in Russia, 1906–1917: Peasant Responses to Stolypin's Project of Rural Transformation.* New York: Oxford University Press.

Skocpol, Theda. 1979. *States and Social Revolutions : A Comparative Analysis of France, Russia, and China.* New York: Cambridge University Press.
Suzuki, Koichi, L. Alexander Norsworthy, and Helen C. Gleason. 1993. *The Clinton Revolution: An Inside Look at the New Administration.* Lanham, Md.: University Press of America.
Suzuki, Koichi, L. Alexander Norsworthy, and Wayne Brough. 1992. *Dawn of a New Era: U.S. Foreign and Domestic Policy at a Crossroads.* Tokyo: NHK Publishing.
Voslensky, Michael. 1984. *Nomenklatura : The Soviet Ruling Class.* 1st ed. Garden City, N.Y.: Doubleday.